ENERGY SCIENCE, ENGINEERING AND TECHNOLOGY SERIES

MICROALGAE FOR BIOFUEL PRODUCTION AND CO_2 SEQUESTRATION

ENERGY SCIENCE, ENGINEERING AND TECHNOLOGY SERIES

Air Conditioning Systems: Performance, Environment and Energy Factors
Tobias Hästesko and Otto Kiljunen (Editors)
2009. ISBN: 978-1-60741-555-8

Oil Shale Developments
Ike S. Bussell (Editor)
2009. ISBN: 978-1-60741-475-9

Power Systems Applications of Graph Theory
Jizhong Zhu
2009. ISBN: 978-1-60741-364-6

Bioethanol: Production, Benefits and Economics
Jason B. Erbaum (Editor)
2009. ISBN: 978-1-60741-697-5

Bioethanol: Production, Benefits and Economics
Jason B. Erbaum (Editor)
2009. ISBN: 978-1-61668-000-8
(Online Book)

Energy Costs, International Developments and New Directions
Leszek Kowalczyk and Jakub Piotrowski (Editors)
2009. ISBN: 978-1-60741-700-2

Handbook of Exergy, Hydrogen Energy and Hydropower Research
Gaston Pélissier and Arthur Calvet (Editors)
2009. ISBN: 978-1-60741-715-6

Computational Techniques: The Multiphase CFD Approach to Fluidization and Green Energy Technologies (includes CD-ROM)
Dimitri Gidaspow and Veeraya Jiradilok
2009. ISBN: 978-1-60876-024-4

Introduction to Power Generation Technologies
Andreas Poullikkas
2009. ISBN: 978-1-60876-472-3

Radial-Bias-Combustion and Central-Fuel-Rich Swirl Pulverized Coal Burners for Wall-Fired Boilers
Zhengqi Li
2009. ISBN: 978-1-60876-455-6

Handbook of Sustainable Energy
W. H. Lee and V. G. Cho (Editors)
2010. ISBN: 978-1-60876-263-7

Bio Resource-Based Energy for Sustainable Societies
K. A. Vogt, D. J. Vogt, M. Shelton, R. Cawston, L. Nackley, J. Scullion, M. Marchand, M. Tulee,T. Colonnese, L. Stephan, S. Doty, R. K. Hagmann, T. C. Geary, T. A. House, I. Nwaneshiudu, P. A. Roads, S. Candelaria, L. L. James, E. Kahn, L.X. Lai, A. M. Lee, S. J. Rigdon and R.M. Theobald
2010. ISBN: 978-1-60876-803-5

Stages of Deployment of Syngas Cleaning Technologies
Filomena Pinto, Rui Neto André, and I. Gulyurtlu
2010. ISBN: 978-1-61668-257-6

Stages of Deployment of Syngas Cleaning Technologies
Filomena Pinto, Rui Neto André, and I. Gulyurtlu
2010. ISBN: 978-1-61668-586-7
(Online Book)

Photovoltaics: Developments, Applications and Impact
Hideki Tanaka and Kiyoshi Yamashita (Editors)
2010. ISBN: 978-1-60876-022-0

Nuclear Fuels: Manufacturing Processes, Forms and Safety
Antoine Lesage and Jérôme Tondreau (Editors)
2010. ISBN: 978-1-60876-326-9

Thin-Film Solar Cells
Abban Sahin and Hakim Kaya (Editors)
2010. ISBN: 978-1-60741-818-4

Wind Turbines: Types, Economics and Development
Gerard Martin and Jeremie Roux (Editors)
2010. ISBN: 978-1-60741-849-8

Advanced Materials and Systems for Energy Conversion: Fundamentals and Applications
Yong X. Gan
2010. ISBN: 978-1-60876-349-8

A Solar Car Primer
Eric F. Thacher
2010. ISBN: 978-1-60876-161-6

A Solar Car Primer
Eric F. Thacher
2010. ISBN: 978-1-6087 382-5
(Online Book)

Jatropha Curcas as a Premier Biofuel: Cost, Growing and Management
Claude Ponterio and Costanza Ferra (Editors)
2010. ISBN: 978-1-60876-003-9

Ethanol Biofuel Production
Bratt P. Haas
2010: ISBN: 978-1-60876-086-2

Biodiesel Handling and Use Guide
Bryan D. O'Connery
(Editor)
2010. ISBN: 978-1-60876-138-8

Cool Power:
Natural Ventilation
Systems in Historic Buildings
Carla Balocco and Giuseppe Grazzini
2010. ISBN: 978-1-60876-129-6

Syngas Generation from
Hydrocarbons and Oxygenates
with Structured Catalysts
Vladislav Sadykov, L. Bobrova,
S. Pavlova, V. Simagina,
L. Makarshin, V. Julian,
R. H. Ross
and Claude Mirodatos
2010. ISBN: 978-1-60876-323-8

Corn Straw and Biomass Blends:
Combustion Characteristics
and NO Formation
Zhengqi Li
2010. ISBN: 978-1-60876-578-2

CFD Modeling and Analysis of
Different Novel Designs of
Air-Breathing Pem Fuel Cells
Maher A.R. Sadiq Al-Baghdadi
2010. ISBN: 978-1-60876-489-1

Direct Methanol Fuel Cells
A. S. Arico, V. Baglio
and V. Antonucci
2010. ISBN: 978-1-60876-865-3

A Sociological Look at Biofuels:
Understanding the
Past/Prospects for the Future
Michael S. Carolan
2010: ISBN: 978-1-60876-708-3

Ejectors and their Usefulness
in the Energy Savings
Latra Boumaraf, André Lallemand
and Philippe Haberschill
2010. ISBN: 978-1-61668-210-1

Sustainable Resilience
of Energy Systems
Naim Hamdia Afgan
2010. ISBN: 978-1-61668-483-9

Sustainable Resilience
of Energy Systems
Naim Hamdia Afgan
2010. ISBN: 978-1-61668-738-0
(Online Book)

Utilisation and Development
of Solar and Wind Resources
Abdeen Mustafa Omer
2010. ISBN: 978-1-61668-238-5

Utilisation and Development
of Solar and Wind Resources
Abdeen Mustafa Omer
2010. ISBN: 978-1-61668-497-6
(Online Book)

Coalbed Natural Gas: Energy
and Environment
K.J. Reddy (Editor)
2010. ISBN: 978-1-61668-036-7

Storage and Reliability of Electricity
Edward T. Glasby (Editor)
2010. ISBN: 978-1-61668-248-4

Geothermal Energy: The Resource Under our Feet
Charles T. Malloy(Editor)
2010. ISBN: 978-1-60741-502-2

Wind Power Market and Economic Trends
Martin W. Urley (Editor)
2010. ISBN: 978-1-61668-243-9

High Temperature Electrolysis in Large-Scale Hydrogen Production
Yu Bo and Xu Jingming
2010. ISBN: 978-1-61668-297-2

High Temperature Electrolysis in Large-Scale Hydrogen Production
Yu Bo and Xu Jingming
2010. ISBN: 978-1-61668-697-0
(Online Book)

Biofuels from Fischer-Tropsch Synthesis
M. Ojeda and S. Rojas (Editors)
2010. ISBN: 978-1-61668-366-5

Biofuels from Fischer-Tropsch Synthesis
M. Ojeda and S. Rojas (Editors)
2010. ISBN: 978-1-61668-820-2
(Online Book)

Transient Diffusion in Nuclear Fuels Processes
Kal Renganathan Sharma
2010. ISBN: 978-1-61668-369-6

Transient Diffusion in Nuclear Fuels Processes
Kal Renganathan Sharma
2010. ISBN: 978-1-61668-735-9
(Online Book)

Coal Combustion Research
Christopher T. Grace (Editor)
2010. ISBN: 978-1-61668-423-5

Coal Combustion Research
Christopher T. Grace (Editor)
2010. ISBN: 978-1-61668-646-8
(Online Book)

Shale Gas Development
Katelyn M. Nash (Editor)
2010. ISBN: 978-1-61668-545-4

Shale Gas Development
Katelyn M. Nash (Editor)
2010. ISBN: 978-1-61668-728-1
(Online Book)

Microalgae for Biofuel Production and CO2 Sequestration
Christopher Q. Lan and Bei Wang
2010. ISBN: 978-1-61668-151-7

ENERGY SCIENCE, ENGINEERING AND TECHNOLOGY SERIES

MICROALGAE FOR BIOFUEL PRODUCTION AND CO_2 SEQUESTRATION

CHRISTOPHER Q. LAN

AND

BEI WANG

Nova Science Publishers, Inc.

New York

LIBRARY OF CONGRESS CATALOGING-IN-PUBLICATION DATA

Microalgae for biofuel production and CO2 sequestration / Bei Wang ... [et al.].
p. cm.
Includes index.
ISBN 978-1-61668-151-7 (hardcover)
1. Microalgae--Biotechnology. 2. Biomass energy. 3. Carbon sequestration. I. Wang, Bei, 1963-
TP248.27.A46M53 2009
662'.88--dc22

2009050565

Published by Nova Science Publishers, Inc. ✦ New York

CONTENTS

PREFACE

Owing to their vast diversity and high growth rate, microalgae offer numerous advantages as the most promising photosynthetic organisms for biofuel production and CO_2 bio-sequestration. Two different processes have been proposed for these purposes: microalgael farming and ocean fertilization. This book focuses primarily on the former while providing a brief introduction to the latter.

Chapter 1 briefly discusses the diversity of microalgae, the concept of photosynthesis, and the microalgael species that have been most studied for biofuel production and CO_2 sequestration in microalgael farming settings;

Chapter 2 discusses the nutritional and environmental requirements of microalgae and the media and cultivation systems commonly used for microalgael farming;

Chapter 3 discusses the use of microalgae for CO_2 sequestration using two different approaches: microalgael farming and ocean fertilization;

Chapter 4 introduces a variety of biofuels to be produced using microalgae as feedstock or cell factories;

Chapter 5, which is co-authored by Courchesne N.M. Dorval, Albert Parisien, Bei Wang and Christopher Q. Lan, discusses the principles and recent developments in the metabolic channelling for enhancing production of lipids, the feedstock for biodiesel production;

Chapter 6 presents a novel approach involving factor grouping and Box-Behnken experimental design for microalgael medium optimization;

Chapter 7, which is co-authored by Yanqun Li, Bei Wang, Nan Wu and Christopher Q. Lan, presents the experimental evidence demonstrating the potential of the green algae *Neochloris oleoabundans* for lipid production.

Chapter 1

MICROALGAE FOR CO_2 SEQUESTION AND BIOFUEL PRODUCTION

1.1. MICROALGAE

Microalgae, defined conventionally for the purpose of this book as all unicellular and simple multi-cellular photosynthetic microorganisms including both prokaryotic microalgae (cyanobacteria) and eukaryotic microalgae, are the most important primary producer of the oceans. They are also widely found in other habitats such as lakes, rivers, ponds, wet lands, deserts and even the north and south poles. It was estimated that there are one to ten million microalgael species on the earth (Bold, 1985) and more than 40,000 species have been identified to date. The vast diversity of microalgael species and their capability of high-efficiency photosynthesis for fast growing, solar energy capturing and CO_2 fixation make them the most promising bio-species for CO_2 bio-sequestration and biofuel production.

The microalgael species that have been identified to date can been classified into 11 divisions (10 eukaryotic microalgae plus the prokaryotic cyanobacteria) according to their photosynthetic pigment composition, biochemical constituents, ultrastructure, and life cycle as listed in Table 1.1. Among these microalgae, six classes of them are of primary importance to biofuel production: diatoms (*alias Bacillariophyceae*, belonging to the division Chrysophyta), green algae (Class Chlorophyceae), golden-brown algae (Class Chrysophyceae), prymnesiophytes (Class Prymnesiophyceae), eustigmatophytes (Class Eustigmatophyceae), and blue-green algae or cyanobacteria (Class Cyanophyceae) (Sheehan et al., 1998).

Table 1.1. Main pigments, storage products, and cell coverings of different divisions of microalgae (Barsanti, 2006)

Division	Pigments					
		Chlorophylls	Phycobilins	Carotenoids	Xanthophylls	Storage Products
Cyanophyta	blue-green algae	A	c-Phycoerythrin c-Phycocyanin Allophycocyanin Phycoerythrocyanin	β-Carotene	Myxoxanthin Zeaxanthin	Cyanophycin (argine and asparagine polymer) Cyanophycean starch (α-1,4-glucan)
Prochlorophyta	blue-green algae	a,b	Absent	β-Carotene	Zeaxanthin	Cyanophycean starch (α-1,4-glucan)
Glaucophyta	green algae	A	c-Phycocyanin Allophycocyanin	β-Carotene	Zeaxanthin	starch (α-1,4-glucan)
Rhodophyta	Red Algae	A	r,b-Phycoerythrin r-Phycocyanin Allophycocyanin	α- and β-Carotene	Lutein	Floridean starch (α-1,4-glucan)
Cryptophyta	blue- green algae	a,c	Phycoerythrin -545 r-Phycocyanin	α-,β-, and ε-Carotene	Alloxanthin	starch (α-1,4-glucan)
Heterokontophyta	Golden or Brown algae	a,c	Absent	α-,β-, and ε-Carotene	Fucoxanthin Violaxanthin	Chrysolaminaran (β-1,3-glucan)
Haptophyta	green algae	a,c	Absent	α- and β-Carotene	Fucoxanthin	Chrysolaminaran (β-1,3-glucan)
Dinophyta	dinoflagellates	a,b,c	Absent	β-Carotene	Peridinin, Fucoxanthin, Diadinoxanthin Dinoxanthin Gyroxanthin	starch (α-1,4-glucan)
Euglenophyta	flagellate	a,b	Absent	β- and γ-Carotene	Diadinoxanthin	Paramylon (β-1,3-glucan)
Chlorarachniophyta	mainly green algae	a,b	Absent	Absent	Lutein Neoxanthin Violaxanthin	Paramylon (β-1,3-glucan)
Chlorophyta	green algae	a,b	Absent	α-,β-, and γ-Carotene	Lutein Prasinoxanthin	starch (α-1,4-glucan)

Diatoms

Diatoms, also called Bacillariophyceae, are a class belonging to division Chrysophyta. The cells of diatoms are golden-brown because of the presence of high level of fucoxanthin, a photosynthetic accessory pigment. Several other xanthophylls are present at lower levels, as well as β-carotene, chlorophyll α and chlorophyll c. The main storage compounds of diatoms are triglycerides (TAGs) and chrysolaminarin, a β-1,3-linked carbohydrate. Diatom cell wall contains substantial quantities of polymerized Si. This unique feature has important implications for media preparation and costs in a commercial production facility, because silicate is a relatively expensive chemical. On the other hand, deficiency of silicate can promote lipid (TAG) accumulation in diatoms. It can be employed to provide a controllable means to induce lipid synthesis in a two-stage production process. Diatoms are the most common and widely distributed groups of microalgae on earth. They dominate the phytoplankton of the oceans and are also commonly found in fresh- and brackish waters.

Green Algae

Green algae, including divisions Prochlorophyta and Chlorophyta, have chlorophyll a and chlorophyll b as photosynthetic pigments. Green algae are believed to be the evolutionary progenitors of higher plants and have received more attention than other groups of algae. *Chlamydomonas reinhardtii* (and closely related species), a member of this group, has been studied extensively. It was the first algae to be genetically transformed. Another genus of green algae that has been studied extensively is Chlorella. Several green algae, for instance, *Neochloris oleoabundans*, are known to be able to accumulate large quantities of lipids and efficient in CO_2 fixation (Li et al. 2008a; Liet al. 2008b), making them attractive candidates for combined CO_2 fixation and biofuel production.

Golden-Brown Algae

Golden-Brown algae include the chrysophytes and the synurophytes. They are similar to diatoms with respect to pigment composition. Some chrysophytes have lightly silicified cell walls. They are found primarily in freshwater habitats. Lipids and chrysolaminarin are the most common carbon storage materials of this group.

Prymnesiophytes

Prymnesiophytes (haptophytes) are primarily marine organisms and account for a substantial proportion of the primary productivity of tropical oceans. Some prymnesiophytes produce algael blooms, which may cause serious problems. Prymnesiophytes are often of a golden-brown color because of the presence of the yellow-brown accessory pigments, diadinoxanthin and fucoxanthin. Lipids and chrysolaminarin are the major storage form of this group of algae.

Eustigmatophytes

This group represents an important component of the "picoplankton", which is comprised of a group of small microalgae with cell size in the range of 2-4μm in diameter. The genus Nannochloropsis is one of the few marine species in this class and is commonly found in the world's oceans. Chlorophyll a is the only chlorophyll present in Eustigmatophyte cells. They contain however several xanthophylls that serve as accessory photosynthetic pigments.

Blue Green Algae (Cyanobacteria)

As mentioned previously, Cyanobacteria are not microalgae but a group of photosynthetic bacteria. They are treated here as microalgae in a broader sense for convenience. Cyanobacteria are prokaryotes that contain no nucleus, no chloroplasts, and have a different gene structure than all other microalgae. There are approximately 2,000 species of cyanobacteria, which have been found in a diversity of different habitats. Some members of this group can assimilate atmospheric N_2 and therefore eliminate the need to provide fixed nitrogen for cell growth. A few commercial facilities have been built for cultivation of cyanobacteria, (e.g., *Spirulina platensis*) for production of health foods and other novel products. No member of this class is known to produce significant quantities of storage lipids.

1.2. PHOTOSYNTHESIS

Photosynthesis is the process photosynthetic species, including microalgae, use to capture light energy to produce glucose and other organic carbons from

CO_2. Light energy is converted in the photosynthetic process to chemical bonding energy stored in cell materials (biomass). Photosynthesis involves two major reaction sequences: the light-dependent reactions (the light reactions) and the light-independent reactions (the dark reactions). In the light reactions, light energy is captured and converted to energy currency, NADPH and ATP. The dark reaction involves a sequence of reactions that fix and reduce inorganic carbon utilizing the ATP and NADPH generated in the light reaction. As shown in equation 1.1, the overall result of photosynthesis is that carbon is converted from CO_2 to carbohydrates, $[CH_2O]_n$, using light energy. The carbohydrates are subsequently converted to other cell materials for cell growth and cell maintenance.

$$n\,CO_2 + nH_2O + \text{light} \xrightarrow[\text{Chlorophyll } a]{} (CH_2O)n + nO_2 \qquad (1.1)$$

1.2.1. The Light Reaction

The light reaction, or more precisely the light-dependent reaction, is the first stage of photosynthesis. In this process light energy is converted to chemical energy in the form of energy-carriers ATP and NADPH. NADPH is also one of the major carriers of the reducing power that is required for the reductive anabolism of cells. There are two types of photosynthesis according to the donors of electrons: oxygenic photosynthesis and anoxygenic photosynthesis. In oxygenic photosynthesis, the electron donor is water, producing molecular oxygen as a by-product. In anoxygenic photosynthesis, various electron donors such as H_2S might be used.

The major mechinary required for light reactions include phototosystem I (PS I), Photosystem II (PS II), the photosynthetic electron tranfer chain (ETC) and the ATP synthase. Photosystems I and II are protein complexes containing light capturing pigments, which are responsible for absorbing light energy. Different pigments contained in cells are important characteristics in the taxonomy of microalgae. As shown in Figure 1.1, the light-dependent reactions begin in PS II, when a particular chlorophyll molecule of PS II absorbs a photon to activate an electron (i.e., the electron attains a higher energy level). Because the high energy state of an electron is very unstable, the electron is then transferred through the ETC. As a result, the electron flows from PS II to PS I, where the electron gets the energy from another photon. In the end, the electron is transferred to $NADP^+$, which serves as the final electron acceptor for the ETC and is reduced to NADPH. As will be discussed more in detail later, one major component of the photosynthetic ETC is cytochrome b6f, which couples with ATP synthase in a process called photophosphorylation to generate ATP molecules.

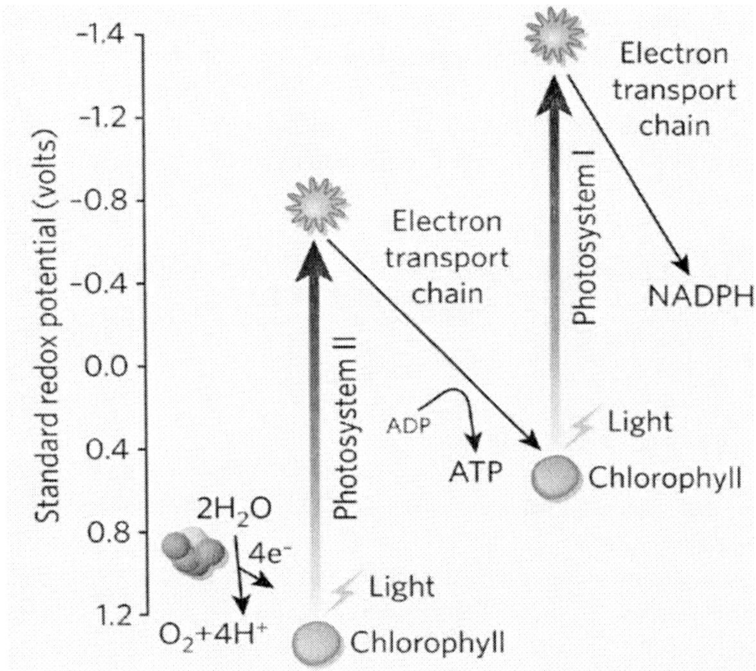

Figure 1.1. The light-dependant reactions of photosysynthesis (Allen & Martin, 2007).

In eukaryotic photosynthetic species such as plants and microalgae, photosynthesis takes place in a specialized organalle called chloroplast. As shown in Figure 1.2, chloroplasts are flat discs usually of 2 to 10 micrometers in diameter and 1 micrometer thick. The chloroplast consists of an inner membrane and an outer membrane, which are separated by the intermembrane space. The material inside chloroplast is called stroma. Chloroplast contains one or more molecules of small circular DNA and some ribosomes. However, most of its proteins are encoded by genes contained in the host cell nucleus and manufactured by ribosomes in the cytosol, with the proteins transported to the chloroplast. Within the stroma are stacks of thylakoids, which are called grana. A thylakoid has a flattened disk shape. Inside a thylakoid is an empty area called the thylakoid space or lumen. While the thylakoid membrane houses all the machinery for the light-dependant reactions and therefore is the location the light reaction takes place, the dark reaction takes place in the stroma of the chloroplast.

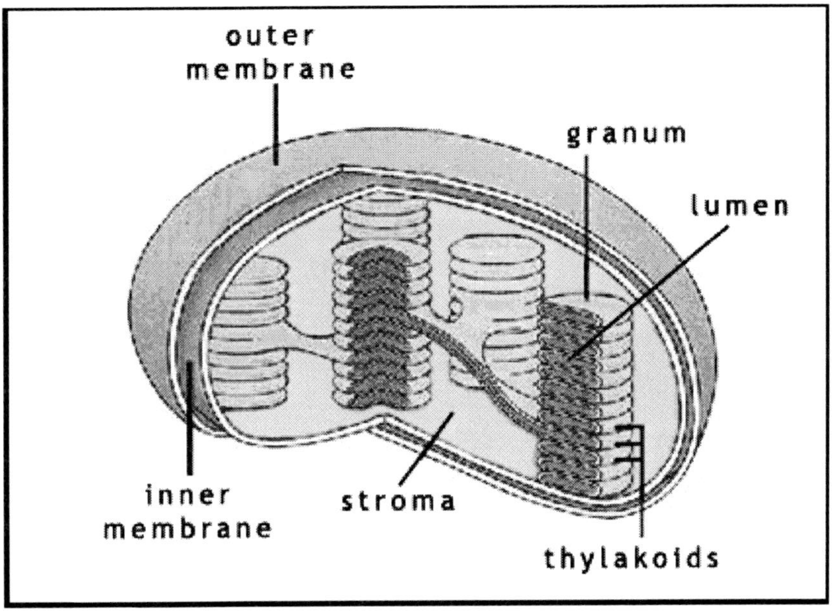

Figure 1.2. *The schematic diagram of the structure of a chloroplas*t. Thylakoids are a phospholipid bilayer membrane-bound compartment. A granum is a stack of thylakoids folded on top of one another. The stroma is the fluid space within the chloroplast. The lumen is the fluid filled space within a thylakoid (image source: http://www.helpsavetheclimate.com/photosynthesis.html).

A more detailed decription of the light-dependant reaction machinary is shown in Figure 1.3, which includes the PS I, the photosynthetic ETC (including plastoquinone (PQ), cytochrome b6f, and plastocyanin (PC)), PS II, and the ATP synthase. In photophosphorylation, the activated electron is first accepted by PQ, the primary electron acceptor. Then, the cytochrome b6f uses the energy of electron to pump protons from the outside of the thylakoid membrane (the chloroplast stroma) to the inside of the thylakoid (the thylakoid lumen) to create a proton gradient across the thylakoid membrane. The proton gradient drives the ATP synthase, which locates across the thylakoid membrane, to form ATP. Photophosphorylation may occur in two different ways: noncyclic photophosphorylation and cyclic photophosphorylation. In non-cyclic photophosphorylation, cytochrome b6f uses the energy of electrons from PSII only and the electron passed to PS I, which is re-activated by the photon absorbed by the PS I complex, is accepted by a $NADP^+$ to form a NADPH molecule. In cyclic photophosphorylation, cytochrome b6f uses the energy of electrons from both PS II and PS I to produce more ATP molecules. No NADPH moleclues are produced in cyclic photophosphorylationi. Cyclic phosphorylation is important to maintain the right ATP/NADPH ratio the light independent reactions.

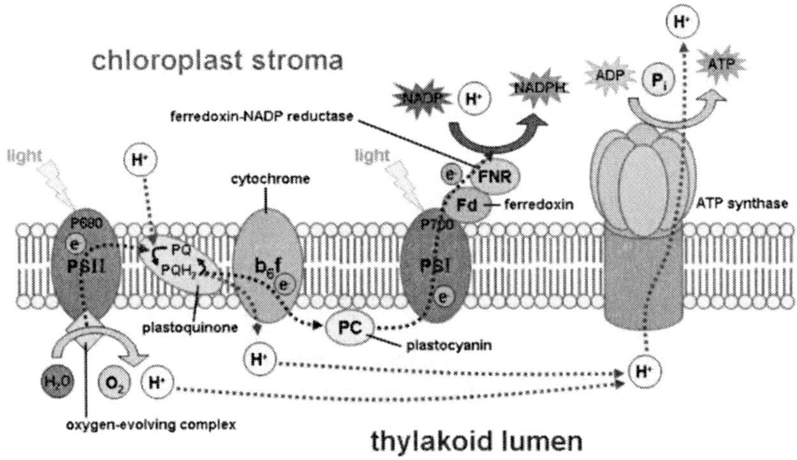

Figure 1.3. thylakoid membrane (Image source: http://en.wikipedia.org/wiki/File:Thylakoid_membrane.png).

Prokaryotic cyanobacteria do not have chloroplast as specialized organelle for photosynthesis. The photosynthetic machinery of most cyanobacteria is embedded into folds of the cell membrane, which is also called thylakoid. Photosynthesis in cyanobacteria is generally oxygenic. However, some may also conduct anoxygenic photosynthesis using hydrogen sulfide as the electron donor. The water-oxidizing oxygenic photosynthesis is accomplished by coupling the activity of PS II and I. They can also use only PS I - cyclic photophosphorylation with electron donors other than water (e.g., H_2S, thiosulphate, or even molecular hydrogen). Their photosynthetic electron transport shares the same compartment as the components of respiratory electron transport. Actually, their plasma membrane contains only components of the respiratory chain, while the thylakoid membrane hosts both respiratory and photosynthetic electron transport.

The overall light-dependent reaction in oxygenic photosynthesis can be expressed by the following equation:

$$12H_2O + 12NADP^+ + 18ADP + 18P_i \rightarrow 6O_2 + 12NADPH + 18ATP \qquad (1.2)$$

1.2.2. Dark Reactions

In the dark reactions, cells fix CO_2 using the ATP and NADPH generated from the light reactions to produce organic compounds (carbohydrates such as glucose). The dark reactions take place in the chloroplast stroma of eukaryotic microalgae and plants and in the cytosol of prokaryotic cyanobacteria. At least four different pathways of dark reactions in different autotrophic organisms, including the Calvin cycle (Calvin-Benson- Bassham cycle or reductive pentose phosphate cycle), the reductive citric acid cycle, the reductive acetyl-CoA pathway, and the 3-hydroxypropinate cycle, have been identified in different photosynthetic organisms. Calvin cycle is the most wide spread CO_2 biofixation pathway among photosynthetic organisms.

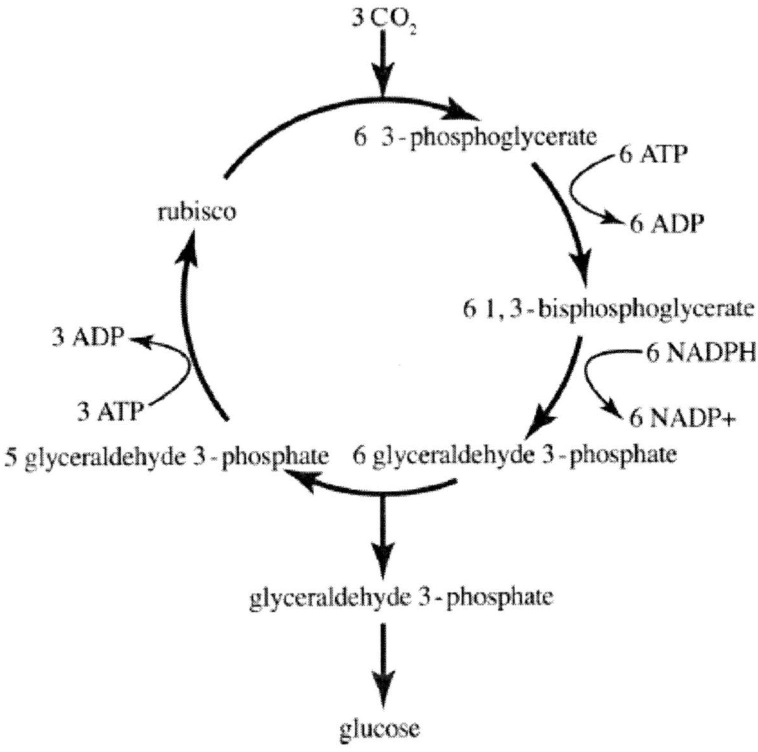

Figure 1.4. the light-independent reaction of photosynthesis: the Calvin cycle.

The Calvin cycle is also called the reductive pentose phosphate cycle. As shown in Figure 1.4, the key step of the Calvin cycle is catalyzed by the enzyme ribulose bisphosphate carboxylase, which fixes a CO_2 molecule onto a molecule of ribulose-1,5-diphosphate (RuBP), resulting in two molecules of glyceric acid-3-phosphate (3PG). These 3PG molecules are then converted into two glyceraldehydes-3-phosphate (G3P, a.k.a. phosphorglyceroldehyde, PGAL) molecules by adding a high-energy phosphate group from ATP to each molecule. The two 3PG molecules are then converted to a RuBP molecule, which stays in the cycle for another around of CO_2, and an organic carbon unit [C]. Three rounds of Calvin cycle lead fixation of 3 CO_2 molecules and the production of a G3P molecule, which can be further converted to glucose, lipids and other cell materials. The overall equation for the Calvin cycle is given by the following equation:

$$3\ CO_2 + 9\ ATP + 6\ NADPH + 6\ H^+ \rightarrow C_3H_6O_3\text{-phosphate} + 9\ ADP + 8\ P_i + 6\ NADP^+ + 3\ H_2O \quad (1.3)$$

1.3. Some Important Species For Biofuel Production

Microalgae contain lipids as membrane components, storage products, metabolites and sources of energy (Rowan, 1989).The microalgael strains with high oil or lipid content are of great interest in the search for a sustainable feedstock for the production of biodiesel. The following species or strains are currently being studied for biofuel production.

1.3.1. *Neochloris oleoabundans*

Neochloris oleoabundans is a microalgae belonging to the class Chlorophyceae. (Fig.1.5) The genus Neochloris belongs to the oily group of microalgae due to its characteristic of producing more than 30% total lipids Containing polyunsaturated fatty acids C18:1(oleic acid) and C18:2 (linoleic acid). In conditions of osmotic shock, the cells excrete high quantities of polysaccharides into the medium and show an increase in the neutral lipids. It has been cultivated in mineral medium deficient in nitrogen and the yield of lipids was 35-54% of cell dry weight. Triglycerides comprised 80% of the total lipids. Aliphatic hydrocarbons, sterols, pigments, glycolipids and phospholipids comprised the remaining lipid fraction. Saturated, monounsaturated and diunsaturated octodecanoic acid represented approximately one-half of the total fatty acids (2-3 weeks, nitrogen starvation).

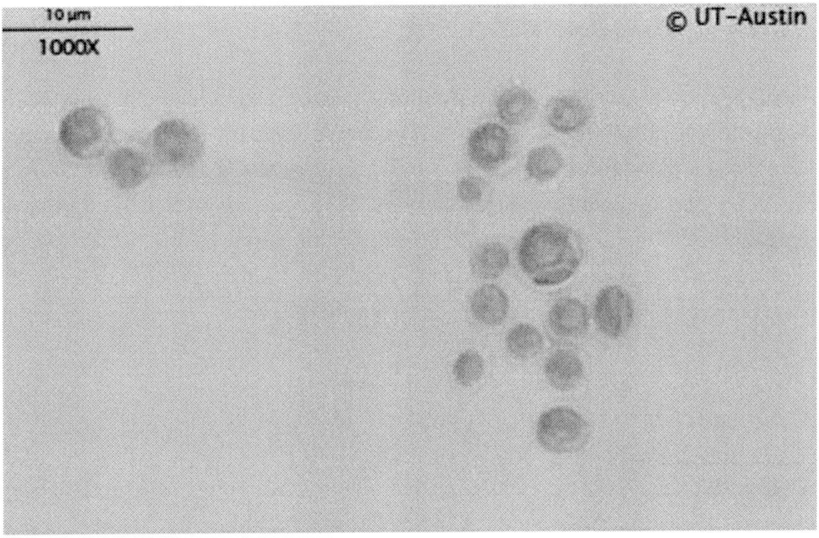

Figure 1.5. *Neochloris oleoabundan.*

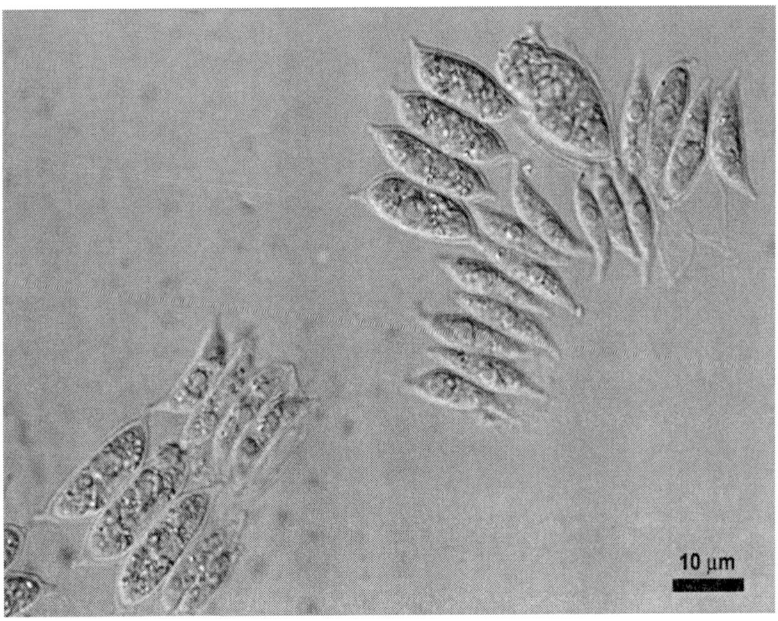

Figure 1.6. *Scenedesmus dimorphus.*

1.3.2. *Scenedesmus dimorphus*

Scenedesmus dimorphus is a unicellular algae in the class Chlorophyceae. It commonly forms colonies of four cells. In the "classical" form the four cells are arranges abreast in which the outer two cells are crescentshaped (facing outwards) and the inner two are parallel pointed ovals. (Fig.1.6) All four cells appear to be joined at the nucleus. (Zvi Cohen, 1999) While this is one of the preferred species for oil yield for biodiesel, one of the problems with *Scenedesmus* is that it's heavy, and forms thick sediments if not kept in constant agitation.

1.3.3. *Nannochloropsis salina*

Nannochloropsis salina is also called *Nannochloris oculata* and belongs to the class of Eustigmatophyceae. A microscopic picture of *Nannochloropsis* is shown in Fig. 1.7. The Eustigmatophyceae can contain the pigments of the xanthophyll cycle-zeaxanthin, antheraxanthin, and violaxanthin. In the same group are *Nannochloris atomus Butcher*, *Nannochloris maculata Butcher*, *Nannochloropsis gaditana Lubian*. Species in the genus *Nannochloropsis* are characterized by unique biochemical and ultrastructure features such as the absence *of chlorophyll b* or *c* and the composition of the cellular xanthophyll pigments, relatively high content of EPA and the presence of specific sterols. *Nanochloropsis* biomass is examined and evaluated as a reliable enriched source of EPA of current applications.

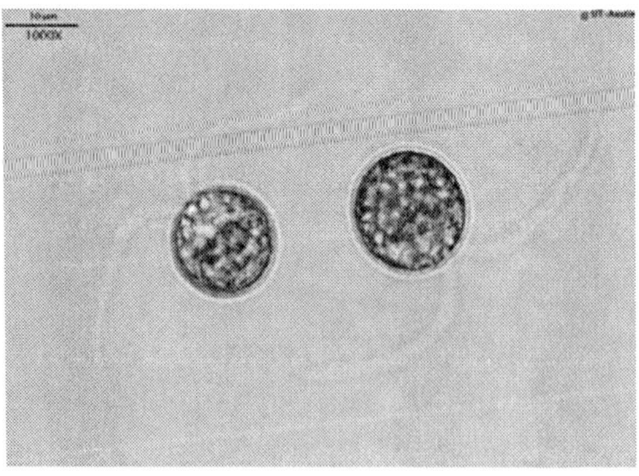

Figure 1.7. *Nannochloropsis salina.*

1.3.4. Botryococcus Braunii

Botryococcus braunii is a colonial member of Chlorophyceae, characterized by a unique organization of colonies and considerable lipids productivity. The colonies appear to consist of single or multiple cell clusters united by transparent and refringent strands. (Fig. 1.8) This species can produce long chain hydrocarbons up to 86% of its dry weight. The green algae *Botryococcus* is unique in the quality and quantity of the liquid hydrocarbons it produces. Some scientists consider the ancestors of *Botryococcus* to be responsible for many of the world's fossil fuel deposits.

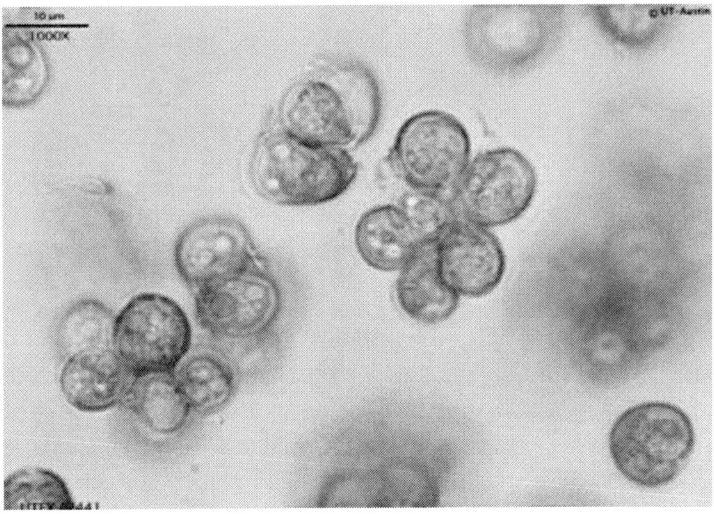

Figure 1.8. *Botryococcus braunii.*

1.3.5. Dunaliella species

The biflagellated algae Dunaliella (Fig. 1.9) is classified under the class Chlorophyceae (order, Volvocales), which includes a variety of ill-defined marine and fresh water unicellular species. Dunaliella occurs in a wide range of marine habitats such as oceans, brine lakes, salt marshes and salt water ditches near the sea, predominantly in water bodies containing more than 2M salt and a high level of magnesium. Unlike other green algae, Dunaliella lacks a rigid polysaccharide cell wall, which permits rapid cell volume changes in response to extracellular changes in osmotic pressure.

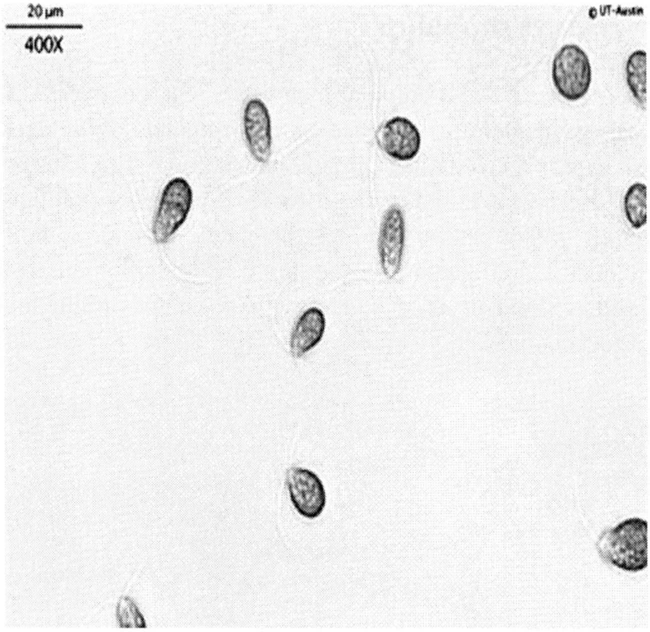

Figure 1.9. *Dunaliella tertiolecta.*

Dunaliella uses a special mechanism of asmoregulation to adapt to the salt concentrations in the surrounding media by varying the intracellular concentration of glycerol in response to the extracellular osmotic pressure. The cellular organization presenting one big chloroplast with single-centred starch surrounded pyrenoid, a few vacules, nucleus and nucleolus, makes this species a potential feedstock for alcohol production, with high growth rates and high starch content. It was also reported that *D. tertiolecta* is capable to accumulate up to 50% lipid on a dry cell weight basis, making it a promising candidate for biodiesel production using seawater. It is a fast growing species, implying great potential for CO_2 bio-sequestration.

1.3.6. Chlorella Species

Chlorella is a genus of unicellular green algae, belonging to the *Phylum Chlorophyta*. It is spherical in shape, about 2 to 10 µm in diameter, and is without flagella (Fig.1.10). *Chlorella* contains the green photosynthetic pigments *chlorophyll a* and *b* in its chloroplast.

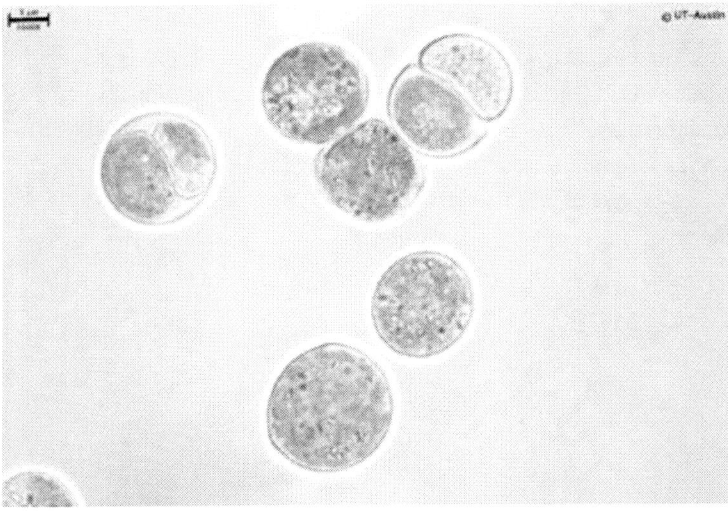

Figure 1.10. *Chlorella sp.*

Chlorella species are in general fast-growing and easy to cultivate. It is well documented that some *Chlorella* species such as *Chlorella minutisimma*, *Chlorella vulgaris* and *Chlorella homosphaera* can accumulate large amounts of fatty acid derivatives, mostly triacylglycerols under controlled conditions. These unique features make them suitable for biofuel production and CO_2 bio-sequestration via microalgael farming.

References

Allen J.F., Martin W. (2007), *Evolutionary biology: Out of thin air, Nature* 445: 610-612.

Barsanti L. and Gualtieri P. (2006), *Algae: Anatomy, biochemistry, and Biotechnology*. CRC, Taylor&Francis Group, London New York.

Bold HC, W. M. (1985). *Introduction to the Algae,* Pretice-Hall, Inc. Englewood Cliffs, NJ, USA.

Becker E.W (1994). *Microalgae: Biotechnology and Microbiology*, Cambridge University Press.

Cohen Z. (1999). *Chemicals from microalgae*, CRC, Taylor&Francis Group, London New York.

Li, Y., M. Horsman, Wang B., Wu N., and Lan C.Q. (2008a). "Effects of nitrogen sources on cell growth and lipid accumulation of green algae Neochloris oleoabundans." *Applied Microbiology and Biotechnology* 81(4): 629-636.

Li, Y., M. Horsman, Wu N, and Lan C.Q., Dubois-Calero N. (2008b). *"Biofuels from microalgae." Biotechnology* Progress 24(4): 815-820.

Rowan K.S. (1989). *Photosynthetic pigments of algae,* Cambridge University Press.

Chapter 2

CULTIVATION OF MICROALGAE

2.1. MICROALGAEL NUTRITION AND MEDIA

Most algael groups are photoautotrophs, i.e., depending entirely upon photosynthesis for cell growth and cell maintenance using sunlight as the source of energy and CO_2 the source of carbon. On the other hand, there are some colorless heterotrophic species that depends on organic carbon sources from the external environment either by taking up dissolved substances (osmotrophy) or by engulfing bacteria and other microorganisms (phagotrophy). A large group of algae can utilize both inorganic and organic carbon sources and is referred to as mixotrophs. Some mixotrophs are primarily photosynthetic and only use organic energy sources occasionally. Others meet most of their nutritional demand by phagotrophy but may use some of the products of photosynthesis from sequestered prey chloroplasts. Of particular importance to CO_2 mitigation and biofuel production are phototrophs and mixtrophs that are primarily photosynthetic.

Hydrogen (H), oxygen (O), carbon (C), nitrogen (N), phosphorus (P), and sulfur (S) are the most important elements constituting algael cells. Growth medium must provide sufficient nutrients for microalgael growth. While hydrogen and oxygen are usually provided in the form of water (H_2O) and molecular oxygen (O_2), other elements, including C, N, P, S, metal ions such as iron, magnesium, and trace elements, and in some cases silicon (Rebolloso-Fuentes, et al. 2001), must be supplied. It is important to develop balanced media for optimal microalgae cultivation and CO_2-fixation (Mandalam and Palsson 1998).

Carbon Sources

According to the mode of cell growth (heterotrophic, autotrophic, or mixotrophic), microalgae can utilize organic and/or inorganic carbon sources for cell growth. From the perspective of microalgae cultivation, the most common organic carbon sources for heterotrophic and mixotrophic cultivation of microalgae are glucose, sucrose, and other sugars derived from starch, sugar cane, lignocellulosic biomass, and other sugar sources. There are three different sources of inorganic carbons sources: 1) CO_2 from the atmosphere; 2) CO_2 from industrial exhaust gases (e.g., flue gas and flaring gas); and 3) fixed CO_2 in the form of soluble carbonates (e.g., $NaHCO_3$ and Na_2CO_3). A more detailed discussion about the fixation of CO_2 from different inorganic carbon sources can be found in "Chapter 3 CO_2 bio-mitigation by microalgae".

Nitrogen Sources

While some microalgael species could fix molecular nitrogen for cell growth, nitrate, ammonia and urea (or combinations of them) are the most common nitrogen sources for microalgae (Lourenco, Barbarino et al. 1998). Worth mentioning is that ammonium is the chemical form of nitrogen most readily taken up and assimilated by phytoplankton. Unlike nitrate, it does not require reduction prior to being assimilated into amino acids. However, research indicated that ammonium at high concentration has toxic effects on microalgael growth (Lourenco, Barbarino et al. 2002). Besides, ammonia may escape into the atmosphere, causing environmental and economic concerns.

A cheap source of nitrogen would be wastewater or secondary wastewater, which contains large quantities of different forms of nitrogen sources. However, the use of wastewater for microalgae cultivation may cause contamination problems and/or complicate downstream processing and therefore should be used with precautions.

It would be necessary for the microalgae to have a high calorific value if they are used for biofuel production through the aforementioned biomass conversion processes. Microalgae grown under normal conditions have been shown to have calorific values between 18 and 21 KJ g^{-1} (Illman, Scragg et al. 2000), which can be improved by optimizing cultivation conditions. For instance, studies have shown that the calorific values of microalgae biomass could be enhanced by cultivation in nitrogen-limiting condition, The calorific value of *C. vulgaris* biomass, which was 18 KJ g^{-1} grown in nitrogen sufficient medium, was found to be increased to 23 KJ g^{-1} grown in low nitrogen medium and the calorific value of *C. emersonii* grown in low nitrogen medium was found to be 29 KJ g^{-1}. Although this calorific value of 29 KJ g^{-1} is somewhat lower than that of diesel, which is 43 KJ g^{-1}, *Chlorella* biomass are still regarded as suitable for use as diesel replacements (Illman, Scragg et al. 2000).

The major components of microalgae biomass are protein, carbohydrate and lipids. Lipids are the most desirable component from the energetic point of view. The cells with a higher lipid content and lower carbohydrates and proteins have elevated calorific value and produce higher yields of oil when processed via, for instance, biomass liquefaction (Tornabene, Holzer et al. 1983; Ginzburg 1993; Yamaberi, Takagi et al. 1998; Illman, Scragg et al. 2000; Scragg, Illman et al. 2002). Several microalgael strains have been reported to have the ability to accumulate large quantities of lipids. Nitrogen limitation was observed to lead to the increase of the lipid content in some chlorella strains such as *C. emersonii* (63%), *C. minutissima* (56%) , C. *vulgaris* (57.9%), *C. luteoviridis* (28.8%), *C. capsulata* (11.4%), and *C. pyrenoidosa* (29.2%) (Illman, Scragg et al. 2000). An oil-rich microalgael species, *Neochloris oleoabundans* (Kawata, Nanba et al. 1998), was reported that under nitrogen deficient conditions, accumulated 35-54% lipids of its cell dry weight and triglycerides comprised 80% of the total lipids (Tornabene, Holzer et al. 1983). It was also observed (Yamaberi, Takagi et al. 1998) that the triglycerides accumulated in *Nannochloris sp.* cells could be 2.2 times as that in the cells in nitrogen sufficient cultures. It is recommended that the intracellular triglyceride content could be increased by prolonging the cultivation period during the stationary phase after nitrogen depletion. Table 2.1 summarizes a few microalgael strains that have been studied for lipid productivity.

However, there is a dilemma in the fact that high-lipid and high caloric cells are usually produced in stress states, which is associated with reduced cell division (Ratledge 2002). Biomass yield and overall lipid/energy productivity will therefore be compromised as a result. For instance, studies on *Chlorella vulgaris* and *C. emersonii* grown in a 230 l pumped tubular photo-bioreactor in Watanabe's medium and a low nitrogen medium (Scragg, Illman et al. 2002) showed that, the low nitrogen medium induced higher lipid accumulation in both algae with calorific value increased and the highest calorific value was obtained with *C. vulgaris* (28 kJ g^{-1}) grown in the low nitrogen medium. However, the biomass productivity was 24 mg dry wt l^{-1} d^{-1} in the low nitrogen medium, only slightly higher than half of that obtained with Watanabe's medium (40 mg dry wt l^{-1} d^{-1}). The overall energy recovery was lower with the low nitrogen medium than with the Watanabe's medium. It is important to find the balance between producing high caloric value cells and maintaining high biomass productivity by optimizing nitrogen in medium for the growth of microalgae.

Table 2.1. Lipid productivity of some microalgael species (Li Y, et al, 2008)

Species	PDCW (g/l)/d	Lipid (%)	TAG (%)	T (oC)	Plipid (mg/l·d)	Refs
Parietochloris incise	(7)/14		43-77	25	17.9	(Solovchenko, Khozin-Goldberg et al. 2008)
Parietochloris incise	(6.3)/38			26	31.5	(Cheng-Wu, Cohen et al. 2002)
Nannochloris sp. UTEX LB1999	(2.7)/12	34.0	18.8	28	76.5	(Yamaberi, Takagi et al. 1998)
Nannochloris sp. UTEX LB1999	(2.16)/	50.9	47.6			(Takagi, Watanabe et al. 2000)
Chlorella protothecoides *	(16.8)/8	57.8	NA		1214	(Xiong, Li et al. 2008)
Chlorella emersonii	(1.11)/14	63	NA		50	(Illman, Scragg et al. 2000)
Chlorella	(0.46)/14	57	NA		18.7	(Illman, Scragg et al. 2000)
Chlorella vulgaris	(0.52)/14	40	NA		14.9	(Illman, Scragg et al. 2000)
Chlorella vulgaris	NA	56.6	NA	20	NA	(Liu, Wang et al. 2008)
Dunaliella	(0.5)/10	67	NA	28	33.5	(Takagi, Karseno et al. 2006)
Neochloris oleoabundans	NA	35-54	80	28	NA	(Tornabene, Holzer et al. 1983)
Neochloris oleoabundans	(2.4)/6	34	NA	34	134	Current Study

* Heterotrophic cultivation

Phosphorus Sources

Phosphorus is another element that has significant relevance to the cell growth and metabolism of microalgae. It is one of the essential elements comprising DNA, RNA, ATP and cell membrane materials, etc. It is worth noting that, as a constituent element of ATP, phosphorus is essential to the cellular processes related to energy transfer (e.g., photophosphorylation). On another relevant notion, photosynthesis requires large amounts of proteins (notably

Rubisco) and the proteins are synthesized by phosphorus-rich ribosomes (Agren 2004).

As aforementioned, phosphorus-containing ATP/ADP are essential for photophosphory lation. As a consequence, limitation of growth by phosphate starvation may have a severe impact on various aspects of microalgael metabolism, including photosynthesis and lipid accumulation.

Phosphorus is preferentially assimilated as inorganic phosphates in the form of $H_2PO_4^-$ and HPO_4^{2-} (Gauthier and Turpin 1997; Martinez, Jimenez et al. 1999). It has been pointed out that phosphates may form complexes or precipitations with some metal ions and not all the added phosphorus is bioavailable (Yun, Lee et al. 1997). Therefore, phosphorus may need to be supplied in excess.

Other Elements

Sulfur, iron, magnesium and other elements are also indispensable for the growth of microalgae. Sulfur is an essential component of cysteine and methionine. In the absence of sulfur, protein biosynthesis is impeded and the photosynthetic system PSII repair cycle is blocked (Zhang, Happe et al. 2002). Magnesium is required for nitrogenase activity using a creatine phosphate/kinase/ATP generating system as one of its roles in cell metabolism. Iron is involved in electron flow from H_2O to $NADP^+$ (Roden and Zachara 1996). Some trace metals play key roles in (non-cyclic) photosynthetic electron transport (Raven, Evans et al. 1999). For instance, manganese is essential for O_2 evolution and calcium has an important role in the thylakoid lumen in facilitating H_2O dehydrogenation and O_2 evolution.

2.2. Other Factors Affecting Microalgael Cell Growth

The most important parameters regulating algael growth by means of photosynthesis, in addition to the previously discussed nutritional factors, are light, pH, turbulence, salinity, and temperature. The range of optimal conditions as well as the toleratable range of operating conditions are species specific and various factors may be interdependent (Laura Barsanti 2006).

Temperature

In general, the cultivation temperature should ideally be as close as possible to the temperature at which the organisms were collected (polar regions <10 °C, temperate regions 10-25 °C, and tropical regions >25 °C). Some species of microalgae tolerate temperatures above 30 °C, which could be ideal candidates for applications such as CO_2 mitigation of flue gas. A temperature between 18-25 °C is most often employed in microalgael farming.

Temperature simultaneously influences three competing cellular processes of microalgae, photosynthesis, photo-respiration, and endogenous metabolism (i.e., cell maintenance). The overall effect of temperature on cell growth in a particular temperature range depends on the net result of these competing processes. Temperature may also influence the production of metabolites as a result of its influence on the metabolism of microalgae.

2.2.2. Light

Light is the source of energy for the autotrophic growth of microalgae. Light intensity, spectral quality, and photoperiod need to be considered to optimize a microalgael farming process. Light intensity plays an important role; however, it depends greatly on the depth and the cell density of the algael culture. At large depths and cell concentrations the light intensity must be increased to penetrate through the culture. Nevertheless, precautions should be taken as too high light intensity (e.g. direct sunlight, small container close to strong artificial light) may result in photo-inhibition. The most commonly employed light intensities range between 100 and 200μE $sec^{-1}m^{-2}$, which corresponds to about 5-10% of full daylight (2000 μE $sec^{-1}m^{-2}$). Moreover, overheating due to both natural and artificial illumination should be avoided. Some microalgael species do not grow well under constant illumination hence a light/dark (LD) cycle may be required (maximum 16:8 LD, usually 14:10 or 12:12).

Light as the energy source is often the principal limiting factor in microalgae cultivation. When the light intensity is below the light saturation point, the rate of photosynthesis is directly correlated to light intensity in a typical scenario. In most microalgae, photosynthesis is saturated at about 30% of the total terrestrial solar radiation, i.e. 1,700–2,000 μE/m^2 s (Kirk, 1994). It should be noticed that light inhibition (i.e., photoinhibition) might occur when light intensity goes beyond certain point. Photoinhibition could be reversible or irreversible, depending on the light stress and the length of time the microalgae are exposed to such a stress. It was reported that microalgae could adjust, within certain range, to a higher light intensity through a process called photo-adaptation.

It is worth mentioning that the natural rethyme of light/dark cycle is incredible for microalgael farming designed for solar energy capturing and CO_2 sequestration. Due to the lack of light energy in nighttime, microalgae would have to conduct respiration, which consumes glucose and/or other cell materials to produce bioenergy (ATP) for the maintenance of cell viability. As the same in any other heterotrohic organisms, respiration of microalgae consumes oxygen and releases CO_2. As a result, significant loss of biomass (up to 30%) could be observed in the nighttime. The respiration during nighttime should therefore be minimized.

Considerable efforts have gone into the design of photobioreactors that would maximize photosynthetic efficiency. It was noted the highest photosynthetic efficiencies recorded to date have been achieved in open ponds and raceways exposed to full sunlight of up to 2000μmol quanta $m^{-2}s^{-1}$. There is considerable scope for design of process improvements that would further enhance this result. One example of process improvement wasthe removal of auto-inhibitory growth factors by medium replacement, which can lead to a four-fold enhancement in area productivity. This particular method could be accomplished by a two-stage cultivation process composing of a first stage carried out in closed photobioreactors and the second stage in open ponds , in which culture medium from the photobioreactor is significantly diluted upon transfer to the second-stage open-pond batch culture (Huntley and Redalje 2007).

pH

The optimal pH range for most algael species is 7 - 9. Some species have pH optima in more acid or basic ranges. For instance, the optimal pH of the cyanobacterium *Spirulina platensis* is in the range of 8.0 - 9.0. It is crucial to maintain culture pH in the optimal range as complete culture collapse due to the disruption of many cellular processes can result from failure to maintain an acceptable pH. In the case of high-density algael culture in controlled systems using air enriched with carbon dioxide (pure CO_2 or high CO_2 flue gases), the balance between the mass transfer of CO_2 from gas phase to liquid phase and the consumption of CO_2 by algael cells will determine the dissolved CO_2 concentration in the culture, which may be the determinant factor of the culture pH in some scenarios.

The CO_2/O_2 Balance

Photosynthesis of microalgae uptakes CO_2 as carbon source and releases O_2 as product. Nevertheless, Ribulose-1,5-bisphosphate carboxylase oxygenase (Rubisco), the prime carboxylating enzyme that furnishes CO_2 for the Calvin cycle can also utilize O_2 for photorespiration. The dissolved CO_2 and O_2 must be balanced to minimize photorespiration. This could be problematic with high cell density algael cultures, in which sufficient CO_2 must be available while evolved O_2 has to be removed before reaching inhibitory concentrations.

In high-cell-density microalgael cultures with optimal growth, species-specific O_2 evolution rates between 28 and 120 mg O_2 g^{-1} DCW h^{-1} were recorded. Oxygen of such a high concentration may become a problem not only because of photorespiration but also the toxicity of oxygen. Upon exposure to strong sun light radiation, oxygen radicals may develop, which are toxic to cells due to membrane damage. Many algael strains cannot survive in significantly O_2-oversaturated environment for longer than 2–3 hours. High temperatures and photon flux density (PFD), combined with CO_2 limitation, will intensify the physiological inhibitory effects. The CO_2/O_2 balance could be improved by 1) increasing turbulence to facilitate gas exchange; 2) stripping O_2 from air; and 3) aerating algael culture using CO_2 enriched air (e.g., using flue gases). All these approached represent some engineering challenges in photobioreactor design, which will be discussed in the following sections.

Salinity

Marine algae are very tolerant to changes in salinity. Most species grow best at a salinity that is slightly lower than that of their native habitat, which is typically obtained by diluting sea water with tap water. Salinities of 20 - 24 g/l are found to be optimal for most marine species. Nevertheless, it was observed by some researchers that exposure of algael cells at abnormally high salinity may result in noticeably high lipid cell content (see discussion in Chapter 5).

Mixing

Microalgae live in their natural habitats at a density of 10^3 cells/ml and at distances of more than 1,000 μm between cells. However, in high-cell-density microalgael cultures, the cell density could be as high as 10^9 cells/ml, creating a microenvironment that differs drastically from what the nature provides and is not optimal for cell growth and productivity. Mixing is therefore necessary to facilitate mass and heat transfer between the microenvironment surrounding individual cells as well as the bulk environment for optimal cell growth and productivity. More specifically, mixing of microalgael cultures is necessary because of the following reasons: 1) to prevent sedimentation of the algae; 2) to ensure that all cells of the population are equally exposed to light and nutrients; 3) to facilitate heat transfer and avoid thermal stratification; 4) to improve gas exchange between the culture medium and the air, which is critical for good CO_2 mass transfer and for avoiding oxygen toxicity. It should be noted that in the ocean cells seldom experience turbulence and hence mixing should be gentle.

Sterility and Species Control

It is well accepted that certain extent of impurity in the cultivation of microalgae cultivation must be tolerated when the processes are designed for low-value objectives such as biofuel production and CO_2 sequestration. Nevertheless, precautions must be taken to avoid excessive contamination. Fortunately, for autotrophic microalgael farming facility, contamination of heterotrophic microorganism such as bacteria is usually not of significant concern due to the lack of organic carbon sources in the system. However, the control of exotic and invasive algael species is critical for stable and continuous operations and also for stable quality of products. Species control could be particularly difficulty for cultivation of relatively slow growing microalgael species.

2.3. Cultivation Systems

Studies have shown that well designed cultivation systems may lead to significant increase of CO_2 fixation efficiency (Javanmardian and Palsson 1991; Kadam 1997; Usui and Ikenouchi 1997). The design of large-scale culture systems have to consider many factors, including light intensity, temperature, biology of the algae, nature of the product, mixing, aeration, source of carbon dioxide and sterilization (Borowitzka 1999), etc. Although microalgae are considered to be relatively efficient for capturing solar energy for the production of organic compounds via photosynthetic process, the photosynthetic efficiency of microalgae for the conversion of solar energy is typically below 20% (Li et al.

2008) . On the other hand, increasing the density of cultures decreases photon availability to individual cells, which reduces specific growth rate of cells. Therefore, the poor penetration of light could be the most significant limiting factor in microalgael cultivation and well designed cultivation systems should be able to address this limitation as much as possible.

There are three major types of cultivation systems, open ponds with moderate surface to volume ratios (3-10 m^{-1}) and photobioreactors with high surface to volume ratios (25-125 m^{-1}) (Weissman, Goebel et al. 1988) (see Table 2.2), and hybrid systems that combine open ponds with photobioreactors.

Table 2.2. Comparison between raceway ponds and tubular photobioreactors of microalgael cultivation (Wang B., et al, 2008)

System	Raceway ponds	Tubular photobioreactors
Light efficiency	Fairly good	Excellent
Temperature control	none	excellent
Gas transfer	poor	low-high
Produced oxygen	low	high
Accumulation	low	low-high
Hydrodynamic stress on algae	difficult	easy
Species control	none	achievable
Sterility	low	high
Cost to scale-up	low	high
Volumetric productivity	high	low

The choice of cultivation systems is the key aspect that significantly affects the efficiency and cost-effectiveness of a microalgael farming process. This topic has been discussed extensively by a few authors (Chaumont 1993; Lee 2001; Pulz 2001; Carvalho, Meireles et al. 2006; Chisti 2007). Carvalho (2006) explained several closed systems in detail. Lee (2001) discussed a few open systems and systematically compared them with closed systems over different geographical regions. Pulz (2001) focused more on process parameters and suggested a large number of open systems. Janssen et al. (2003) offered useful conceptual diagrams for some of the discussed closed systems and described new systems to be examined, including the use of optical fibres to enhance lighting. Even though the open pond systems seem to be favoured for commercial cultivation of microalgae at present due to their low capital costs, closed systems offer better control over contamination, mass transfer, and other cultivation conditions. The following is a brief summary of different systems available for microalgae cultivation.

2.3.1. Natural Habitats

Microalgae are found in a large diversity of different natural habitats such as oceans, lakes, rivers, and ponds. These natural systems could be employed for microalgael farming. Oceans are the largest natural habits of microalgae and are responsible for approximately 50% of primary production on the earth. As will be discussed in Chapter 3, ocean fertilization is regarded as the most promising strategy for CO_2 bio-sequestration.

Other natural systems such as eutrophic lakes or small natural basins can also be exploited for microalgael production, provided suitable climatic conditions and sufficient nutrients are available. Barsanti and Gualtieri (2006) described a few examples of microalgael farming using natural water bodies, which include the numerous temporary or permanent waters along the northeast border of Lake Chad, where *Arthrospira sp.* grows almost as monoculture and has been collected as human foods by the Kanembou people in those regions. *Arthrospira sp.* naturally blooms also in old volcanic craters filled with alkaline waters in the Myanmar region. Production began at Twin Taung Lake in 1988, which, by 1999, has reached to approximately 100 tons per year. About 60% is harvested from boats on the season in the summer, when the cyanobacterium forms thick mats on the lake. People in boats collect a dense concentration of *spirulina* in buckets. *Arthrospira* is harvested on parallel inclined filters, washed with fresh water, dewatered, and pressed again. This paste is extruded into noodle like filaments which are dried in the sun on transparent plastic sheets. Similar to lakes, natural ponds that do not necessitate mixing and need only minimal environmental control represent another type of extensive employed cultivation systems.

2.3.2. Open Ponds

Different types of artificial open ponds, which vary in size, shape, materials used for construction, and mixing device, have been designed and experimented with for microalgae cultivation. Large outdoor ponds can be unlined with a natural bottom or lined with materials such as clay, brick, cement, or plastic sheets, glass fiber. While lined ponds requires relatively high capital costs, unlined ponds suffer from silt suspension, percolation, and potentially heavy contamination with their applications limited to a few algael species and particular soil and environmental conditions.

The most common design of artificial open ponds (Vonshak and Richmond 1988) are raceway ponds (see Figs. 2.1 & 2.2), in which algael cultures are mixed in a turbulent flow sustained by a paddle wheel. At least two types of open raceway ponds have been used commercially. The first is raceway ponds lined by concrete and the second is a shallow earthen tunnel lined with PVC or other durable plastic. The size of commercial ponds varies from 0.1 ha to 0.5 ha.

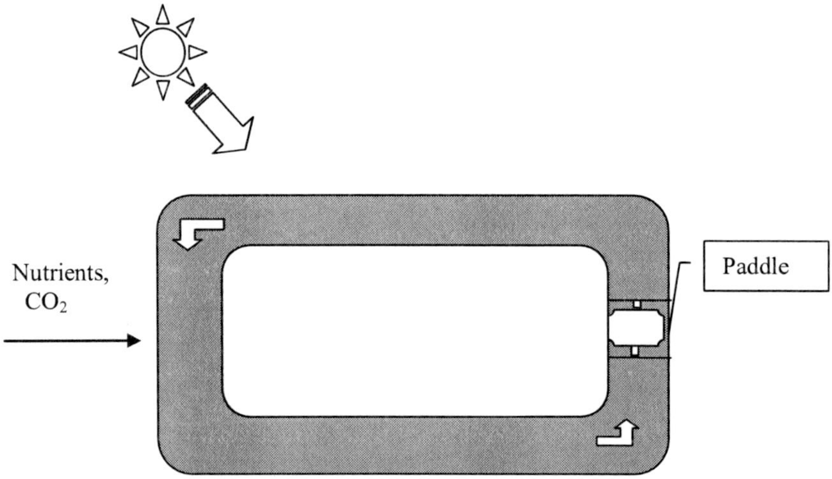

Figure 2.1. A single raceway open pond unit.

Figure 2.2. A microalgael farm using raceway ponds (Earthrise Farms, California, USA).

2.3.3. Closed Systems: Photobioreactors

An alternative to open ponds for large-scale production of microalgael biomass are photobioreactors, which are closed systems that do not allow direct gas exchange between the algael culture and the atmosphere. While open ponds require relatively low capital costs, photobioreactors provide other advantages (Rosello Sastre, Csogor et al. 2007) such as 1) large surface/volume ratios; 2) ability to prevent contamination (species control); 3) excellent control of operating condictions; and 4) capacity to achieve high density of biomass (high biomass productivity and therefore high CO_2 fixation rate, see Table 2.2). These devices provide a protected environment for cultivated species, relatively safe from contamination by other microorganisms. In a photobioreactor, cultivation conditions such as pH, dissolved oxygen concentration (DO) and carbon dioxide concentration (DCO_2) and temperature can be better controlled. Moreover, the water loss through evaporation can be minimized and higher CO_2 utilization/sequestration efficiency could be achieved. The capacity of closed photobioreactors to maintain high cell concentration implies high system productivity and relatively low downstream processing costs. However, these systems are more expensive to build and operate.

2.3.4. The Hybrid Photobioreactor/Open Pond Cultivation System

So far, industrial-scale cultivation has been limited to open pond technology and was successful only for a handful of microalgael species (Borowitzka 1999). Cultivation in open ponds of microalgae, although carefully selected for their high oil content, as a feedstock for biodiesel fuel is still unsustainable at present stage. Nevertheless, closed-systems are well recognized for their excellent ability to control sterility and, in so doing, to permit continuous cultivation of a large variety of species in a monoculture manner, although their application to industrial production has been limited to small-scale processes, generally less than 1000 L. The primary problem associated with the use of photobioreactors is the aforementioned large capital cost (Terry and Raymond 1985), which was estimated to be approximately US \$100 per m^2. Although it is expected that significant cost improvements are bound to occur as the technology is further developed, it is still, at present stage, a popular view that "cost constraints restrict consideration of microalgael biodiesel production systems to the simplest possible devices, which are large unlined, open, mixed raceway ponds" (Sheehan et al. 1998).

The apparent technical dilemma is that the open pond technology is affordable at industrial scale but cannot provide long-term production due to difficulties in species control, while photobioreactors provide good species control but are restrained by high capital and operational costs. The hybrid photobioreactor/pond cultivation system, in couple with a two-stage strategy microalgael cultivation approach, offers a possible solution to this problem.

As will be discussed in detail in Chapter 5, limitation of nutrients such as nitrogen has often been used to force the cessation of cell growth of microalgae and direct the metabolic flux generated from photosynthesis to lipid production. Since lipids are extracellular products, the lipid productivity depends on two parameters, the cell density and lipid cell content. Since cells overproduce lipids grow slowly due to metabolic burden, it is advantages to divide the whole process into two stages: cell growth stage in which cells are cultivated in nutrient rich medium to achieve high biomass productivity, and then the cells are suspended in nutrient-limiting medium so that individual cells, although stop division, could accumulate large quantities of lipids. The two-stage approach was demonstrated to be able to work well when coupled with a hybrid system algael oil production using *Haematococcus pluvialis* (Huntley and Redalje 2007). Closed photobioreactors were used for the cell growth stage and open ponds for the lipid production stage. In the first stage, rich media containing all the required nutrients for the cell growth of the microalgael species were used to achieve fast cell growth and the employment of tightly controlled closed photobioreactors allowed good species control. In the second stage, nitrogen limiting medium was used in open ponds for elevated lipid accumulation. Species control was achieved relatively easily in the second stage because: 1) the cell density of the cultivated microalgael species was high in the second stage system; and 2) depletion of nitrogen in the medium for lipid accumulation prevented the growth of wildtype algae in open ponds. In this demonstration project, a commercial-scale (2 ha) demonstration microalgael farming facility was operated consecutively for 4 years to produce *H. pluvialis* for biodiesel production. Daily production of 1.9 kg dry biomass was achieved with a 25,000 L photobioreactor, corresponding to a biomass productivity of 0.076 g l^{-1} d^{-1}, at a biomass concentration of 0.3 g/l. An annual averaged rate of microalgael oil production, which was equivalent to 420 GJ ha^{-1} yr^{-1}, was obtained. While the maximum production rate achieved with *H. pluvialis* was equivalent to 1014 GJ ha^{-1} yr^{-1}, it was predicted that a rate of 3200 GJ ha^{-1} yr^{-1} is feasible using fast-growing *Chlorella* species. This is a rate possible to replace the reliance on current fossil fuel usage equivalent to about 300 EJ yr^{-1} and eliminate fossil fuel emissions of CO_2 of about 6.5 Gigatons of Carbon (GtC) per year using only 7.3% of the surplus arable land projected to be available by

2050. It was also expected that other microalgael biodiesel processes such as the one being developed at the University of Utah would be cost-competitive with regular diesel by 2009 (Seefeldt 2007). There is no doubt that global efforts from both the public and private sectors will be continued and accelerated in order to make biofuels from microalgae a practical replacement of fossil fuels in the near future.

2.4. Design of Photobioreactor

As pointed out by Zijffers et al. (2008), an ideal photobioreactor for production of biomass should catch all the sunlight available at the allocated spot, and transport, channel, and distribute it in such a way into the cultivation vessel that all caught light energy is used for biomass formation. In other words, high yields of biomass can only be achieved by linking photobioreactor design to the biological processes inside. The efficiency of photobioreactor is determined by the integration of light capturing, light transportation, light distribution, and light usage.

A well designed photobioreactor should allow convenient and precise control on parameters that have key effects on the productivity of microalgae. These parameters, as discussed in section 2.2, include temperature, light, pH, dissolved oxygen, dissolved CO_2, mixing, and sterility and species control. The following general principles therefore should be observed to enable the photobioreactors the ability to achieve this objective.

2.4.1. Choice of Materials and Dimensions

The material used for construction of photobioreactors must have good light transmission to allow maximum light penetration and the thickness of the wall should provide sufficient mechanical strength of the selected material. The dimensions of a photobioreactor vary according to the design of individual photobioreactor type but should maximize the volumetric productivity of the photobioreactor.

It is important to keep in mind that in natural habitats microalgae are exposed to natural sunshine without any obstacle and the maximum cell density is approximately 10^3 cells/ml. On the other hand, high cell densities of up to 10^9 cells/ml are expected in a photobioreactor for enhanced productivity. The high cell density would drastically reduce the transmission of light through cell suspension. Consequently, selection of high transparent material for the construction of photobioreactors is very important to allow the capture of solar energy as much as possible.

2.4.2. Gas Exchange: Aeration and Mixing

Good gas exchange is of critical importance because, as discussed before, low dissolved CO_2 could limit algael cell growth while high dissolved O_2 concentration is inhibitory. A photobioreactor should include the following feature for gas exchange: 1) space for gas exchange, either as part of the body of a photobioreactor or as a detached vessel; and 2) ability to maintain an appropriate level of turbulenceby the means of aeration, pumping, agitation, or the combination of them.

2.4.3. Temperature Control

The nature of microalgael farming, i.e., capturing solar energy for CO_2 fixation and cell growth, implies that a commercial facility will most likely be located outdoors and be exposed to a large range of day/nigh and seasonal temperature changes. Furthermore, the surface of cultivation system, no matter it is open pond or photobioreactor, has to be exposed to sunshine as much as possible to maximize solar energy receiving. Devising cost-effective and reliable temperature control strategy is therefore a significant challenge of photobioreactor design.

2.4.4. Mixing

Depending on the scale and the choice of the cultivation system, mixing is achieved by stirring daily by hand (test-tubes) shaker (Erlenmeyer flasks), aerating (bags and tanks), paddle wheels and jet pumps (ponds), agitators (stirred tank photobioreactor), and turbulent liquid and air flows (tubular and airlift photobioreactors). It should be noticed that not all algael species can tolerate vigorous mixing.

2.4.5. Cleanability

Cleanability is of critical importance to a photobioreactor for the following reasons: 1) prevent formation of biofilm on the wall and therefore maintain high light transmission; and 2) minimize the chance of contamination. To increase the cleanability, the following principles should be followed: 1) the internal surface of a photobioreactor should be smooth and free-drop; 2) minimize the number of internals; 3) in the case of tubular photobioreactor, minimize the number of bends; 4) the internal diameter of a photobioreactor should be large enough to allow convenient cleaning.

2.4.6. Prevention of Biofilm Formation

Formation of biofilm on the internal surface of photobioreactor may drastically reduce the light transmission and the sterility of the photobioreactor, resulting in poor productivity and stability of operation. To minimize biofilm formation, the following precautions should be taken: 1) choose materials that can prevent the adhesion of algae onto the surface; 2) the photobioreactor should be able to maintain enough turbulence in operation (e.g., by means of pumping, agitation, aeration) to minimize biofilm formation; 3) superb cleanibility of the internal surface of photobioreactor.

2.5 Typical Photobioreactors

A variety of different photobioreactors have been developed. These photobioreactors could be classified in several ways according to different criteria such as the structure of photobioreactor (tubular, column or flat panel); the layout of photobioreactor (horizontal, inclined, vertical, or spiral); the means of mixing power delivery (compressed air, pumping, or agitator) and the means of illumination (external or internal illumination, conventional or illuminated with optical fibres). We will discuss the design of photobioreactors according to their structures, i.e., tubular, column, and flat-panel photobioreactors in this section.

2.5.1. Tubular Photobioreactors

Tubular photobioreactor is one of the most popular configurations of photobioreactors (Travieso, Hall et al. 2001). It typically includes an array of transparent tubes built in different patterns (straight, bent, or spiral). Relatively small tube diameter, generally 0.1 m or less, is necessary for ensuring high biomass productivity. Tubular photobioreactors could have different orientations including horizontal, inclined, vertical arrangements.

As shown in Figure 2.3, a tubular photobioreactor cultivation is comprised of the following components: 1) the solar array for algae growth, 2) the harvesting unit to separate algae from the suspension, 3) a degassing column for gas exchange and cooling (heating) and 4) a circulation pump. Fresh medium is fed into the degassing column. A pilot microalgael farm employing an array of tubular photobioreactors is shown in Figure 2.4.

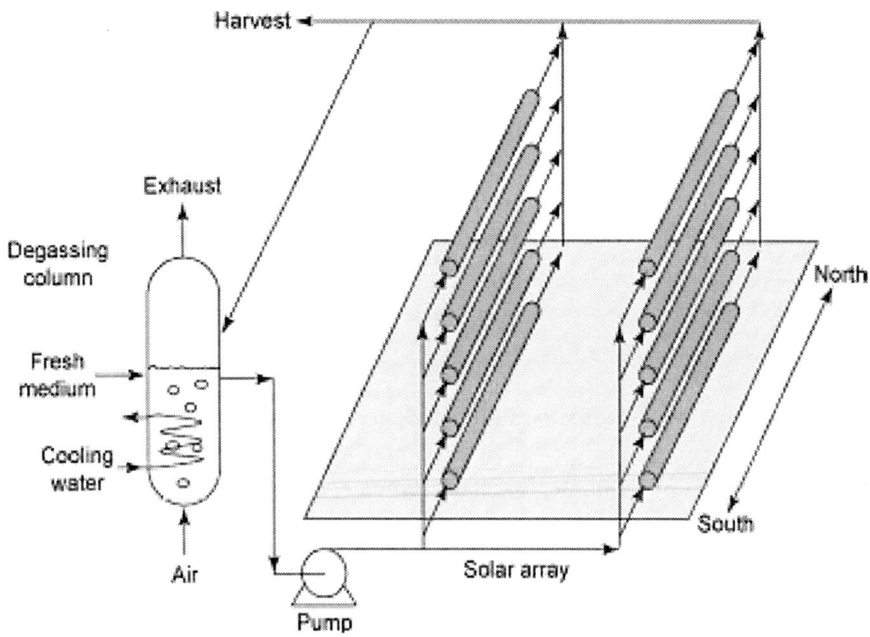

Figure 2.3. Schematic diagram of a tubular photobioreactor cultivation system (Chisti, Y., 2008).

Figure. 2.4. Pilot tubular photobioreactor: horizontal arrangement (www.70centsagallon.com/Algae.html).

It should be cautioned that excessive supply of radiant energy to narrow tubular PBR may lead to photon loss or even photoinhibition, especially at low-cell-density phase. This problem could be partial overcome by using tubular PBR of large diameter coupled with improved mixing. Light inhibition may also be lessened by shading. For instance, it was demonstrated with *Chlorella sorokiniana* in inclined outdoor tubular photobioreactors equipped with static mixers (Ugwu, Ogbonna et al. 2005) that, when the tube diameter of the photobioreactor was increased from 3.8 cm to 12.5 cm, the volumetric productivities decreased but the areal productivities increased. The effectiveness of the static mixers in improving the volumetric productivity was about 63% higher in large diameter tubular photobioreactors (12.5 cm diameter tubes) than in the small diameter tubular photobioreactors (3.8 cm diameter tubes). The static mixers were also more effective at higher standing biomass concentrations than at low standing biomass concentrations. Installation of static mixers in the tubular photobioreactor resulted in improved biomass yield from solar radiation, partly due to better light distribution among the cells as the cells were moved efficiently between the upper and lower parts of the tubes, and partly due to lower dissolved oxygen concentrations (DO). It was also demonstrated recently that shading of the tubular PBR surfaces to diminish solar irradiance by 70 % led to higher biomass productivity and greater accumulation of total chlorophyll and carotenoids compared to the values obtained when the PBR was completely exposed to full sunlight (Ugwu and Aoyagi 2008).

2.5.2. Flat Panel Photobioreactors

As shown in Figure 2.5, there are two basic types of flat-panel photobioreactors (FP-PBR), pump-driven and airlift FP-PBR, according the means of mixing. Pump-driven FP-PBR depends on the flow of liquid created by pumping to create the necessary turbulence for mixing while airlift FP-PBR depends on compressed air to deliver the power of mixing. Similar to tubular PBR, flat panel (also called flat plate) PBR features large surface/volume ratio and may become a standard reactor type for mass production of several algael species. The investigation of flat plate reactors for microalgae cultivation dates back to the early 1950s (Burlew 1953) and a variety of different flat panel PBR have been developed since then. For instance, Samon and Leduy (1985) proposed a vertically translucent flat plate PBR, which was illuminated on both sides and stirred by aeration. Tredici et al. (Tredici, Carlozzi et al. 1991; Tredici and Materassi 1992) proposed a rigid alveolar panel PBR. Pulz et al. (1995) proposed a flat-panel PBR with inner walls arranged to promote turbulence by pumping. The Pulz reactor as shown in Fig. 2.5, was comprised of parallel plates packed together, which were compact enough to attain 6 m^3 of culture volume on 100 m^2 of ground area, with a total illuminated culture surface of approximately 500 m^2. Hu et al. (1996) proposed a simple method of constructing flat-panel PBR using glass sheets glued together with silicon rubber, enabling easy construction of reactors with any desired light-path. Recently, a new design of vertical flat panel photobioreactor consisting of a disposable plastic bag located between two iron frames has been proposed (Sierra, Acién et al. 2008), bringing about a substantial cost reduction to PBR construction. The disposability of the PBR construction materials makes biofouling less a concern. Nevertheless, the disposal of used plastic bags may present a significant challenge at large scale operations. Most recently, a novel flat panel airlift photobioreactor (FP-ALPBR) was proposed as an alternative system for the cultivation of *Haematococcus pluvialis* NIES-144. The 17-l FP-ALPBR system was capable of giving reasonable growth characteristics with a maximum cell density of 4.1×10^5 cell ml^{-1} and specific growth rate of 0.52 day^{-1} being achieved. A similar level of performance was obtained from a 90-l FP-ALPBR system, i.e., cell density = 40×10^4 cell ml^{-1} but with a slight decrease in specific growth rate to 0.39 day^{-1} (Issarapayup, Powtongsook et al. 2009). The schematic diagram of an airlift flat-panel photobioreactor is shown in Fig. 2.5.

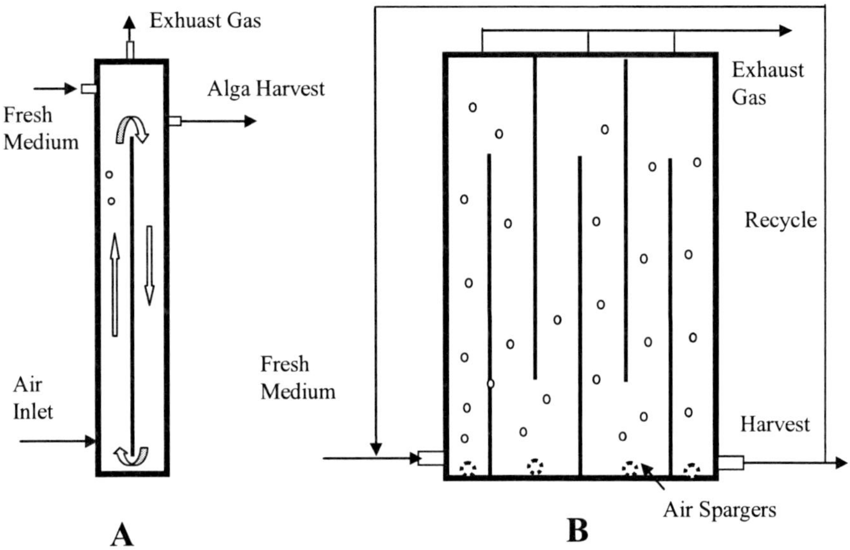

Figure 2.5. Schematic diagrams of two different flat panel photobioreactors: (A) the side view of a single unit of airlift flat-panel PBR; (B) the front view of a pump driven flat-panel photobioreactor.

2.5.3. Cylindrical Column Photobioreactors

The most common cylindrical column photobioreactors are bubble columns and airlift photobioreactors. Airlift photobioreactors include internal-loop airlift and external-loop airlift photobioreactors.

Bubble Column Photobioreactors

Bubble sparged photobioreactors (Berberoglu, et al. 2007) are common choices for microalgae cultivation. As shown in Figure 2.6A, a bubble column PBR is comprised of a column with air sparger located at the bottom of it. The freeboard regime at the top of the PBR is for gas liquid separation. Mixing is achieved by the turbulence created by the air bubbles moving upward. Depends on the design, perforate plates by be installed inside the column to break up bubbles and increase turbulence.

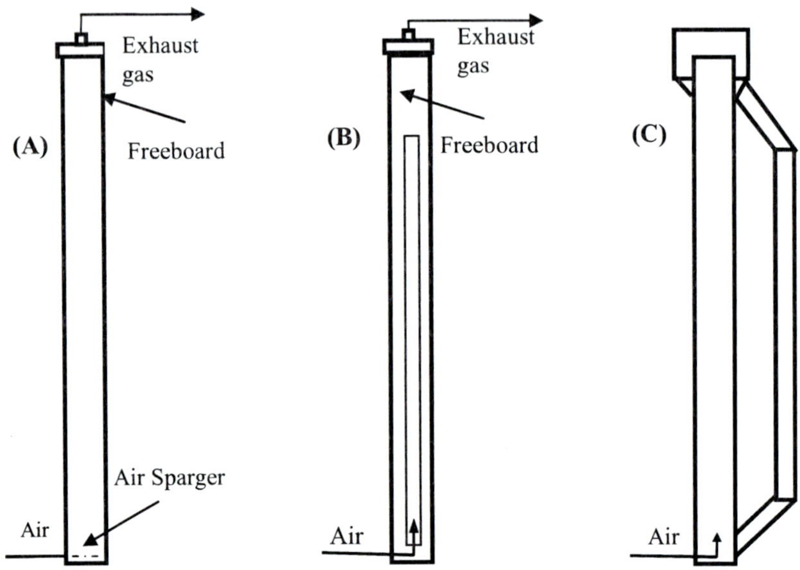

Figure 2.6. schematic diagrams of: (A), a bubble column PBR; (B), an internal-loop airlift PBR; and (C), an external-loop airlift PBR.

External Loop Airlift and Internal Loop Airlift Photobioreactor

The schematic diagram of a conventional internal loop airlift PBR is shown in Figure 2.6 B. It is typically comprised of a transparent column, an internal column, and an air sparger. Air or CO_2 enriched air is introduced inside the internal column; and degassing occurs in the freeboard regime, which locates on top of the internal column. Since the gas holdup inside the internal column is much larger than that in the degassed liquid outside of the internal column, an upward flow of the liquid/gas mixture will be created inside the internal column while a downward flow of degassed liquid is generated outside of it. It is clear that the largest advantage of an airlift PBR is the excellent mixing if offers, which allows good exposure of cells to light radiation even with a relatively large diameter of column and high cell density. On the other side, remarkable advantages of a bubble column include its simplicity and cleanability.

External Loop Airlift Photobioreactor

One of the possible configurations of external airlift PBR is shown in Figure 2.6 C. In an external loop airlift PBR, degassing occurs in a gas/liquid separation on top of the column and circulation of degassed liquid is achieved through an external circulation column. Depending on the design, the circulation of liquid could be facilitated by an external pump.

Performance of Different Bubble Column and Airlift Photobioreactors

A large quantity of reports are available regarding the evaluation of different bubble column and/or airlift photobioreactors. For instance, Oncel and Sukan (2008) compared a internal loop airlift and bubble column PBR with respect to their performances during cultivation of *Artrospira platensis* (*Spirulina platensis*). Culture conditions were kept the same and different parameters were examined through the experiments. It was observed that a higher dry biomass weight and chlorophyll-a concentration was obtained in the airlift PBR, yielding a maximum growth rate of 0.45 day^{-1}, while 0.33 day^{-1} was reached in the bubble column PBR. Subsequently, a 17-day of production was carried out in the selected PBR to fully determine the performance of the PBR. Maximum growth rate of 0.47 day^{-1} was reached during long term cultivation (Oncel and Sukan 2008). Another Relatively large (0.19m column diameter, 2m tall, $0.06m^3$ working volume) outdoor bubble column and airlift bioreactors (a split-cylinder and a draft-tube airlift device) were compared for monoseptic fed-batch culture of the microalgae *Phaeodactylum tricornutum*. These three photobioreactors produced similar biomass versus time profiles and final biomass concentration (~ $4kgm^{-3}$). The maximum specific growth rate, observed within a daily illuminated period in the exponential growth phase, had a value of $0.08h^{-1}$ on the third day of culture. Because of night-time losses of biomass, the specific growth rate averaged over the 4-days of exponential phase was $0.021h^{-1}$ for the three reactors. The biomass in the vertical column reactors did not experience photoinhibition under conditions that are known to cause photoinhibition in conventional thin-tube horizontal loop reactors (photosynthetically active daily averaged irradiance value of 1150±52 $\mu Em^{-2}s^{-1}$). Because of good gas-liquid mass transfer, the dissolved oxygen concentration in the reactors at peak photosynthesis remained less than 120% of air saturation. As a result, oxygen inhibition of photosynthesis and photo-oxidation of the biomass did not occur. More recently, Jacob-Lopes et al. (2009) evaluated bubble column and airlift photobioreactors under three different operational conditions: simple operation, air recirculation and two sequential reactors, to treat air containing 15% carbon dioxide using the cyanobacteria, *Aphanothece microscopica* Nägeli. The results showed that the two-stage sequential photobioreactors (elimination capacity and removal efficiency of 12,217 $g_{carbon}/m^3{}_{reactor}$ day and 52.5%, respectively) to be the operational mode with greatest potential for application on an industrial scale by the increased removal efficiency. .

2.5.4. Photobioreactor Internally Illuminated with Optical Fibre

A variety of different photobioreactors internally illuminated using optical fibres have also been proposed. Internally illuminated PBR could take the structure of any of the aforementioned conventional externally illuminated PBR, however, with solar radiation energy collected using special a panel and transferred to the culture suspension internally using bundles of optical fibers. This type of PBR offers two significant advantages over the externally illuminated PBRs: 1) vastly large illumination area for a given ground area, which is determined by the total surface area provided by the optical fibres rather than the PBR surface area; and 2) in-door cultivation system, which is more robust to day/night and seasonal temperature changes, is made possible. However, capital costs and the cleanibility, which are associated with the complexity of this type of PBRs due to the use of optical fibre, are two major concerns impeding their commercial application at present. As an example, Lee et al. (Lee, Kwon et al. 2005) studied biological CO_2 fixation by *Chlorella* sp. HA-1 in a semi-continuous and series reactor system using an internally illuminated photobioreactor An average CO_2 fixation rate of 4.013 g CO_2 day^{-1} was achieved in a series operation of four reactors.

2.5.5. Membrane Photobioreactors

Membrane photobioreactors employ the large surface areas provided by membranes to facilitate mass transfer while avoiding excessive turbulence, which is energy consuming and sometimes cell damaging, or to separate extracellular metabolites continuously to allow long stable production periods. Different configurations of membrane photobioreactors have been investigated for different purposes. For instance, a photobioreactor coupled with an ultrafiltration system (immersed membranes) was investigated for the continuous cultivation of the microalgae *Haslea ostrearia* in order to improve pigment (marennine) production and recovery. The system presents a commercial interest because the energetic costs were minimized and the cells were not submitted to any shear stress due to pumping or circulation. To obtain this, a membrane module was placed at the bottom of the cylindrical photobioreactor and the hydrostatic pressure (the height of the water column) was used as driving force both for the permeation and periodical backflushing steps. The production of biomass and marennine was stable for a three-week period, with a marennine concentration ~3 times higher than in a conventional batch photobioreactor. This simple photobioreactor–ultrafiltration-with-immersed-membrane concept has been demonstrated of good potentials in biotechnology and aquaculture for continuous extraction of exocellular metabolites (Rossignol, Lebeau et al. 2000). On a different notion, a membrane-sparged helical tubular photobioreactor (MSTR) with a cultivation volume of 800 ml was investigated by Fan et al. (2008)..The A helical tube was used to ensure good light regime, and hollow fiber membranes were uniformly fitted inside the reactor, which functioned as a gas sparger and produced small bubbles. Mass transfer coefficients, mixing intensities and capabilities of CO_2 biofixation through the photosynthesis of *Chlorella vulgaris* in MSTR under different gas, liquid flow rates and light intensities were compared with two other photobioreactors (BCTR and MCTR). BCTR took a perforated pipe as sparger, while MCTR employed a membrane contactor as the whole mass transfer system.

REFERENCES

Agren, G. I. (2004). "The C:N:P stoichiometry of autotrophs - Theory and observations." *Ecology Letters* 7(3): 185-191.

Berberoglu, H., J. Yin, et al. (2007). "Light transfer in bubble sparged photobioreactors for H_2 production and CO_2 mitigation." *International Journal of Hydrogen Energy* 32(13): 2273-2285.

Borowitzka, M. A. (1999). "Commercial production of microalgae: ponds, tanks, tubes and fermenters." *Journal of Biotechnology* 70(1-3): 313-321.

Burlew, J. S. (1953). *Algael Culture from Laboratory to Pilot Plant*: 235-281.

Carvalho, A. P., L. A. Meireles, et al. (2006). "Microalgael reactors: A review of enclosed system designs and performances." *Biotechnology Progress* 22(6): 1490-1506.

Chaumont, D. (1993). "Biotechnology of algael biomass production: A review of systems for outdoor mass culture." *Journal of Applied Phycology* 5(6): 593-604.

Cheng-Wu, Z., Z. Cohen, et al. (2002). "Characterization of growth and arachidonic acid production of Parietochloris incisa comb. nov

(Trebouxiophyceae, Chlorophyta)." *Journal of Applied Phycology* 14(6): 453-460.

Chisti, Y. (2007). "Biodiesel from microalgae." *Biotechnology Advances* 25(3): 294-306.

Chisti, Y. (2008). "Biodiesel from microalgae beats bioethanol." *Trends in Biotechnology* 26(3): 126-131

Fan, L. H., Y. T. Zhang, et al. (2008). "Evaluation of a membrane-sparged helical tubular photobioreactor for carbon dioxide biofixation by Chlorella vulgaris." *Journal of Membrane Science* 325(1): 336-345.

Gauthier, D. A. and D. H. Turpin (1997). "Interactions between inorganic phosphate (P(i)) assimilation, photosynthesis and respiration in the P(i)-limited green algae Selenastrum minutum." *Plant, Cell and Environment* 20(1): 12-24.

Ginzburg, B. Z. (1993). "Liquid fuel (oil) from halophilic algae: a renewable source of non- polluting energy." *Renewable Energy* 3(2-3): 249-252.

Hu, Q., H. Guterman, et al. (1996). "A flat inclined modular photobioreactor for outdoor mass cultivation of photoautotrophs." *Biotechnology and Bioengineering* 51(1): 51-60.

Huntley, M. E. and D. G. Redalje (2007). "CO2 mitigation and renewable oil from photosynthetic microbes: A new appraisal." *Mitigation and Adaptation Strategies for Global Change* 12(4): 573-608.

Illman, A. M., A. H. Scragg, et al. (2000). "Increase in Chlorella strains calorific values when grown in low nitrogen medium." *Enzyme and Microbial Technology* 27(8): 631-635.

Issarapayup, K., S. Powtongsook, et al. (2009). "Flat panel airlift photobioreactors for cultivation of vegetative cells of microalgae Haematococcus pluvialis." *Journal of Biotechnology* 142(3-4): 227-232.

Jacob-Lopes, E., S. Revah, et al. (2009). "Development of operational strategies to remove carbon dioxide in photobioreactors." *Chemical Engineering Journal* 153(1-3): 120-126.

Janssen, M., J. Tramper, et al. (2003). "Enclosed outdoor photobioreactors: Light regime, photosynthetic efficiency, scale-up, and future prospects." *Biotechnology and Bioengineering* 81(2): 193-210.

Javanmardian, M. and B.O. Palsson (1991). "High-density photoautotrophic algal cultures: Design, construction, and operation of a novel photobioreactor system." *Biotechnology and Bioengineering* 38(10): 1182-1189.

Kadam, K. L. (1997). "Power plant flue gas as a source of CO2 for microalgae cultivation: Economic impact of different process options." *Energy Conversion and Management* 38(SUPPL. 1): S505-S510.

Kaewpintong, K., A. Shotipruk, et al. (2007). "Photoautotrophic high-density cultivation of vegetative cells of Haematococcus pluvialis in airlift bioreactor." *Bioresource Technology* 98(2): 288-295.

Kawata, M., M. Nanba, et al. (1998). "Isolation and characterization of a green algae Neochloris sp. for CO2 fixation." *Studies in Surface Science and Catalysis* 114: 637-640.

Barsanti L. and Gualtieri P. (2006). *Algae:Anatomy Biochemistry and Biotechnology,* Taylor & Francis Group. Boca Raton London New York.

Lee, J. Y., T. S. Kwon, et al. (2005). "Biological fixation of CO2 by Chlorella sp. HA-1 in a semi-continuous and series reactor system." *Journal of Microbiology and Biotechnology* 15(3): 461-465.

Lee, Y. K. (2001). "Microalgael mass culture systems and methods: Their limitation and potential." *Journal of Applied Phycology* 13(4): 307-315.

Li Y., Horsman M, Wu N, and Lan C.Q., Dubois-Calero N., *Biofuels from Microalgae,* Biotech Prog, 24: 815-820 (2008)

Li Y., Wang B., Wu N., and Lan C.Q., *Effects of nitrogen sources on cell growth and lipid production of Neochloris oleoabundans*, Appl Microbiol Biotechnol, 81 (4), pp. 629-636 (2008)

Liu, Z. Y., G. C. Wang, et al. (2008). "Effect of iron on growth and lipid accumulation in Chlorella vulgaris." *Bioresource Technology* 99(11): 4717-4722.

Loubière, K., E. Olivo, et al. (2009). "A new photobioreactor for continuous microalgael production in hatcheries based on external-loop airlift and swirling flow." *Biotechnology and Bioengineering* 102(1): 132-147.

Lourenco, S. O., E. Barbarino, et al. (1998). "Distribution of intracellular nitrogen in marine microalgae: Basis for the calculation of specific nitrogen-to-protein conversion factors." *Journal of Phycology* 34(5): 798-811.

Lourenco, S. O., E. Barbarino, et al. (2002). "Effects of different nitrogen sources on the growth and biochemical profile of 10 marine microalgae in batch culture: An evaluation for aquaculture." *Phycologia* 41(2): 158-168.

Mandalam, R. K. and B. Palsson (1998). "Elemental balancing of biomass and medium composition enhances growth capacity in high-density Chlorella vulgaris cultures." *Biotechnology and Bioengineering* 59(5): 605-611.

Martinez, M. E., J. M. Jimenez, et al. (1999). "Influence of phosphorus concentration and temperature on growth and phosphorus uptake by the microalgae Scenedesmus obliquus." *Bioresource Technology* 67(3): 233-240.

Oncel, S. and F. V. Sukan (2008). "Comparison of two different pneumatically mixed column photobioreactors for the cultivation of Artrospira platensis (Spirulina platensis)." *Bioresource Technology* 99(11): 4755-4760.

Pulz, O. (2001). "Photobioreactors: Production systems for phototrophic microorganisms." *Applied Microbiology and Biotechnology* 57(3): 287-293.

Pulz, O., N. Gerbsch, et al. (1995). "Light energy supply in plate-type and light diffusing optical fiber bioreactors." *Journal of Applied Phycology* 7(2): 145-149.

Ratledge, C. (2002). "Regulation of lipid accumulation in oleaginous micro-organisms." *Biochemical Society Transactions* 30(6): 1047-1050.

Raven, J. A., M. C. W. Evans, et al. (1999). "The role of trace metals in photosynthetic electron transport in O2-evolving organisms." *Photosynthesis Research* 60(2-3): 111-149.

Rebolloso-Fuentes, M. M., A. Navarro-Perez, et al. (2001). "Biomass nutrient profiles of the microalgae Nannochloropsis." *Journal of Agricultural and Food Chemistry* 49(6): 2966-2972.

Roden, E. E. and J. M. Zachara (1996). "Microbial reduction of crystalline iron(III) oxides: Influence of oxide surface area and potential for cell growth." *Environmental Science and Technology* 30(5): 1618-1628.

Rossignol, N., T. Lebeau, et al. (2000). "Comparison of two membrane - Photobioreactors, with free or immobilized cells, for the production of pigments by a marine diatom." *Bioprocess Engineering* 23(5): 495-501.

Samson, R. and A. Leduy (1985). "Multistage Continuous Cultivation of Blue-Green Algae Spirulina Maxima in the Flat Tank Photobioreactors with Recycle." *Canadian Journal of Chemical Engineering* 63(1): 105-112.

Scragg, A. H., A. M. Illman, et al. (2002). "Growth of microalgae with increased calorific values in a tubular bioreactor." *Biomass and Bioenergy* 23(1): 67-73.

Seefeldt, L. C. (2007). "Utah group plans to make biodiesel from algae." *Industrial Bioprocessing* 29(3): 5-6.

Sierra, E., F. G. Acién, et al. (2008). "Characterization of a flat plate photobioreactor for the production of microalgae." *Chemical Engineering Journal* 138(1-3): 136-147.

Solovchenko, A. E., I. Khozin-Goldberg, et al. (2008). "Effects of light intensity and nitrogen starvation on growth, total fatty acids and arachidonic acid in the green microalgae Parietochloris incisa." *Journal of Applied Phycology* 20(3): 245-251.

Takagi, M., Karseno, et al. (2006). "Effect of salt concentration on intracellular accumulation of lipids and triacylglyceride in marine microalgae Dunaliella cells." *Journal of Bioscience and Bioengineering* 101(3): 223-226.

Takagi, M., K. Watanabe, et al. (2000). "Limited feeding of potassium nitrate for intracellular lipid and triglyceride accumulation of Nannochloris sp. UTEX LB1999." *Applied Microbiology and Biotechnology* 54(1): 112-117.

Terry, K. L. and L. P. Raymond (1985). "System design for the autotrophic production of microalgae." *Enzyme and Microbial Technology* 7(10): 474-487.

Tornabene, T. G., G. Holzer, et al. (1983). "Lipid Composition of the Nitrogen Starved Green Algae Neochloris Oleoabundans." *Enzyme and Microbial Technology* 5(6): 435-440.

Travieso, L., D. O. Hall, et al. (2001). "A helical tubular photobioreactor producing Spirulina in a semicontinuous mode." *International Biodeterioration and Biodegradation* 47(3): 151-155.

Tredici, M. R., P. Carlozzi, et al. (1991). "A vertical alveolar panel (VAP) for outdoor mass cultivation of microalgae and cyanobacteria." *Bioresource Technology* 38(2-3): 153-159.

Tredici, M. R. and R. Materassi (1992). "From open ponds to vertical alveolar panels: the Italian experience in the development of reactors for the mass cultivation of phototrophic microorganisms." *Journal of Applied Phycology* 4(3): 221-231.

Ugwu, C. U. and H. Aoyagi (2008). "Influence of shading inclined tubular photobioreactor surfaces on biomass productivity of C. sorokiniana." *Photosynthetica* 46(2): 283-285.

Ugwu, C. U., J. C. Ogbonna, et al. (2005). "Characterization of light utilization and biomass yields of Chlorella sorokiniana in inclined outdoor tubular photobioreactors equipped with static mixers." *Process Biochemistry* 40(11): 3406-3411.

Usui, N. and M. Ikenouchi (1997). "The biological CO_2 fixation and utilization project by RITE(1): Highly-effective photobioreactor system." *Energy Conversion and Management* 38(SUPPL. 1): S487-S492.

Wang B., Li Y., Wu N., and Lan C.Q., CO_2 Bio-Mitigation Using Microalgae, *App Microbiol Biotech,* 79: 707-718 (2008)

Weissman, J. C., R. P. Goebel, et al. (1988). "Photobioreactor Design Mixing, Carbon Utilization, and Oxygen Accumulation." *Biotechnology and Bioengineering* 31(4): 336-344.

Xiong, W., X. Li, et al. (2008). "High-density fermentation of microalgae Chlorella protothecoides in bioreactor for microbio-diesel production." *Applied Microbiology and Biotechnology* 78(1): 29-36.

Yamaberi, K., M. Takagi, et al. (1998). "Nitrogen depletion for intracellular triglyceride accumulation to enhance liquefaction yield of marine microalgael cells into a fuel oil." *Journal of Marine Biotechnology* 6(1): 44-48.

Yun, Y. S., S. B. Lee, et al. (1997). "Carbon dioxide fixation by algael cultivation using wastewater nutrients." *Journal of Chemical Technology and Biotechnology* 69(4): 451-455.

Zhang, L., T. Happe, et al. (2002). "Biochemical and morphological characterization of sulfur-deprived and H2-producing Chlamydomonas reinhardtii (green algae)." *Planta* 214(4): 552-561.

Zijffers, J. W. F., M. Janssen, et al. (2008). "Design process of an area-efficient photobioreactor." *Marine Biotechnology* 10(4): 404-415.

APPENDIX: TYPICAL MEDIA FOR MICROALGAE

Bold 3n Medium

Reagent	per liter
$NaNO_3$	0.75 g
$CaCl_2 \cdot 2H_2O$	0.025 g
$MgSO_4 \cdot 7H_2O$	0.074 g
K_2HPO_4	0.075 g
KH_2PO_4	0.175 g
NaCl	0.025 g
Microelement stock solution	6ml

Microelement Stock Solution

$Na_2EDTA \cdot 2H_2O$	0.75 g/L
$FeCl_3 \cdot 6H_2O$	0.097 g/L
$MnCl_2 \cdot 4H_2O$	0.041 g/L
$ZnCl_2$	0.005 g/L
$CoCl_2 \cdot 6H_2O$	0.002 g/L
$Na_2MoO_4 \cdot 2H_2O$	0.004 g/L

Chu Medium

Reagent	per liter
$CaCl_2 \cdot 2H_2O$	0.037 g
$NaNO_3$	0.085 g
$MgSO_4 \cdot 7H_2O$	0.037 g
K_2HPO_4	0.0087 g
$Na_2SiO_3 \cdot 9H_2O$	0.0284 g
Citric Acid·H_2O	0.0335 g
Ferric citrate	0.0335 g
Micronutrient Stock Solution	1ml

Chu Micronutrient Stock Solution

$CuSO_4 \cdot 5H_2O$	0.02 g/L
$ZnSO_4 \cdot 7H_2O$	0.044 g/L
$CoCl_2 \cdot 6H_2O$	0.02 g/L
$MnCl_2 \cdot 4H_2O$	0.012 g/L
$Na_2MoO_4 \cdot 2H_2O$	0.012 g/L
H_3BO_3	0.62 g/L
$Na_2EDTA \cdot 2H_2O$	0.05 g/L
$NaHCO_3$	0.155 g/L

F/2 Medium

Reagent	**Per Liter Seawater**
$NaNO_3$	0.075 g
$NaH_2PO_4 \cdot H_2O$	0.037 g
Microelement stock solution	1 ml
Vitamin solution	1 ml

Microelement Stock Solution

$FeCl_3 \cdot 6H_2O$	3.150 g/L
Na_2EDTA	4.160g
$MnCl_2 \cdot 4H_2O$	0.180 g/L
$CoCl_2 \cdot 6H_2O$	0.010 g/L
$CuSO_4 \cdot 5H_2O$	0.01 g/L
$ZnSO_4 \cdot 7H_2O$	0.022 g/L
$Na_2MoO_4 \cdot 2H_2O$	0.006 g/L

Vitamin Solution

Biotin (Vitamin H)	0.5 mg/L
Thiamine HCl (Vitamin B_1)	100mg/L
Cyanocobalamin (Vitamin B_{12})	0.5mg/L
pH=adjust to 8.0 with 1M NaOH or HCl	0.5mg/L

Allen Medium

Reagents	**Per liter**
HEPES buffer	2.3 g
$NaNO_3$	1.5 g
K_2HPO_4	0.0375 g
$MgSO_4 \cdot 7H_2O$	0.0375 g
Na_2CO_3	0.02 g
$CaCl_2 \cdot 2H_2O$	0.025 g
$Na_2SiO_3 \cdot 9H_2O$	0.058g
Citric Acid $\cdot H_2O$	0.006 g
P-IV Metal Stock Solution	1 ml

P-IV Metal Stock Solution

$Na_2EDTA \cdot 2H_2O$	0.75 g/L
$FeCl_3 \cdot 6H_2O$	0.097 g/L
$MnCl_2 \cdot 4H_2O$	0.041 g/L
$ZnCl_2$	0.005 g/L
$CoCl_2 \cdot 6H_2O$	0.002 g/L
$Na_2MoO_4 \cdot 2H_2O$	0.004 g/L

MES-Volvox Medium

Reagent	Per liter
$Ca(NO_3)_2 \cdot 4H_2O$	0.118 g
Na_2glycerophosphate$\cdot 5H_2O$	0.05 g
KCl	0.05 g
MES	1.95 g
NH_4Cl	0.0267 g
P-IV Metal Stock Solution	6 ml
Vitamin B_{12}	1ml
Biotin Vitamin Stock Solution	1ml

P-IV Metal Stock Solution

$Na_2EDTA \cdot 2H_2O$	0.75 g/L
$FeCl_3 \cdot 6H_2O$	0.097 g/L
$MnCl_2 \cdot 4H_2O$	0.041 g/L
$ZnCl_2$	0.005 g/L
$CoCl_2 \cdot 6H_2O$	0.002 g/L
$Na_2MoO_4 \cdot 2H_2O$	0.004 g/L

Vitamin B_{12}

HEPES buffer pH 7.8	2.4 g/200 ml dH_2O
Vitamin B_{12}(cyanocobalamin)	0.027 g/200 ml dH_2O

Biotin Vitamin Solution

HEPES buffer pH 7.8	2.4 g/200 ml dH_2O
Biotin	0.005 g/200 ml dH_2O

Chapter 3

CO_2 BIO-MITIGATION BY MICROALGAE

3.1. INTRODUCTION

The total CO_2 emissions from fossil fuels were 6.79 GtC yr^{-1} in 2000 and was projected to increase to 8.35 GtC yr^{-1} in 2010 and 9.97 GtC yr^{-1} in 2020 (Huntley and Redalje 2007). Accelerating emission of CO_2, the primary greenhouse gas (GHG), will likely lead to dramatic changes in the Earth's climate system (Shi and Shen 2003; Wang, Li et al. 2008). Development of cost-effective and sustainable CO_2 fixation strategies has therefore become a focus of extensive research and the principal goal of international environmental policy.

Various CO_2 mitigation strategies have been investigated, which can be generally classified into two categories 1) chemical reaction-based approaches and 2) biological CO_2 mitigation. A popular chemical reaction-based CO_2 mitigation approach is achieved by cyclic carbonation/de-carbonation reactions in which gaseous CO_2 reacts with solid metal oxide (represented by MO) to yield metal carbonate (MCO_3) (Gupta and Fan 2002). The reaction can be represented by eq. 3.1.

$$MO + CO_2 \longrightarrow MCO_3 \qquad (3.1)$$

Once the metal oxide reaches its ultimate conversion, metal carbonate can be thermally regenerated to metal oxide and CO_2 by heating the metal carbonate beyond its calcination temperature. The calcination reaction can be represented by Eq. 3.2. An actual installation of this chemical reaction-based CO_2 separation process in a fossil-fuel-fired utility would consist of a carbonation reactor and a regeneration reactor. Lime, with CaO being the primary component, is probably the most popular solid adsorbent for chemical reaction-based CO_2 mitigation due to its relatively low costs (Cooper and Alley 1994).

$$MCO_3 \longrightarrow MO + CO_2 \quad (3.2)$$

Another gas-absorption process for the separation of carbon dioxide from a gas mixture is washing with aqueous amine solution (Resnik, Yeh et al. 2004), among which monoethanolamine (Bonenfant, Mimeault et al.) is the most widely employed (Blauwhoff, Versteeg et al. 1984). Packed bed or plate columns have been investigated to facilitate the contact of a liquid adsorbent with the gas stream (Lin, Liu et al. 2003; Krumdieck, Wallace et al. 2008). CO_2 desorption and therefore the regeneration of monoethanolamine could be performed by heating the product solution to facilitate the reverse reaction. Water vapor in the regenerated CO_2 could be easily separated by condensation. CO_2 fixation and desorption reactions can be represented by eq. 3.3 and eq. 3.4, respectively.

$$CO_2 + 2C_2H_4OH\text{-}NH_2 + H_2O \longrightarrow C_2H_4OH\text{-}NHCO_2^- + C_2H_4OH\text{-}NH_3^+ \quad (3.3)$$

$$C_2H_4OH\text{-}NHCO_2^- + C_2H_4OH\text{-}NH_3^+ \longrightarrow CO_2 + 2C_2H_4OH\text{-}NH_2 + H_2O \quad (3.4)$$

The chemical reaction-based CO_2 fixation schemes mentioned above typically consist of three procedures: separation, transportation, and sequestration. The cost of CO_2 separation and compression to 110 bar (for transportation) is estimated to be \$30-50 per ton of CO_2, and transportation and sequestration are estimated to cost about \$1-3 per ton per 100 km and \$1-3 per ton of CO_2, respectively (Gupta and Fan 2002; Shi and Shen 2003) Since these methods for capturing CO_2 are relatively costly and energy-consuming, the mitigation benefits become marginal. It is therefore necessary to develop cost-effective and sustainable alternatives to curb the soaring emission.

Biological CO_2 mitigation has attracted much attention as an alternative strategy because it leads to production of biomass energy in process of CO_2 fixation through photosynthesis (Ragauskas, Williams et al. 2006; Kondili and Kaldellis 2007) (Ragauskas, Williams et al. 2006). It was estimated (IPCC 1995) that biological mitigation options could offset 10-20% of projected fossil fuel emissions by 2050.

Biological CO_2 mitigation can be carried out by plants and photosynthetic microorganisms. However, the potential for increased CO_2 capture in agriculture by plants has been estimated to contribute only 3-6% of fossil fuel emissions (Skjanes, Lindblad et al. 2007), largely due to the slow growth rates of conventional terrestrial plants. On the other hand, microalgae, a group of fast growing unicellular or simple multicellular microorganisms, have the ability to fix CO_2 while capturing solar energy with an efficiency 10 to 50 times greater than that of terrestrial plants (Usui and Ikenouchi 1997; Li Y. 2008).

There are two ways to enhance CO_2 sequestration by microalgae: 1) ocean fertilization, which is an approach to enhance the biological production, and therefore the CO_2 sequestrating capacity, of oceans by means of enriching oceans with limiting nutrients such as iron; and 2) large-scale cultivation of microalgae in controlled environments, a process known as microalgael farming. The marine phytoplankton, which is comprised mostly of unicellular microalgae, offers approximately half of the current annual biological CO_2 fixation (approximately 50 billion tons of carbon per annum). Ocean fertilization, which is expected to significantly enhance the bioproductivity of the oceans, is believed by many as the only means at our disposal to slow down the global warming before it gets out of control. Microalgael farming is also expected to play an increasingly important role in CO2 biomitigation.

3.2. MICROALGAEL CO_2 FIXATION PATHWAYS

CO_2 biofixation using microalgae involves the cultivation of microalgae, which capture CO_2 from the atmosphere or other sources (e.g., flue gases) for cell growth and extracellular product synthesis, in different cultivation systems such as natural waters, open ponds, and photobioreactors. In this process, autotrophic microorganisms capture CO_2 to synthesize cell materials and extracellular products. As a result, CO_2 is fixed in the form of organic biomass, which can be converted to bulk chemicals and/or biofuels through biorefinery. The CO_2 biofixation reaction can be represented by the follow equation:

$$nCO_2 + nH_2O \underset{\text{Energy}}{\Longleftarrow\!\!=\!\!=\!\!=\!\!\Longrightarrow} [CH_2O]_n \text{ (biomass)} \quad (3.5)$$

The Calvin cycle, which is also called the reductive pentose phosphate cycle, is the most widespread CO_2 biofixation pathway among microalgae. As shown in Figure 3.2, the key step of the Calvin cycle is catalyzed by the enzyme ribulose bisphosphate carboxylase, which fixes a CO_2 molecule onto a molecule of ribulose-1,5-diphosphate (RuBP), resulting in two molecules of glyceric acid-3-phosphate (3PG). These 3PG molecules are then converted into two glyceraldehydes-3-phosphate (G3P, a.k.a. phosphoglyceraldehyde, PGAL) molecules by adding a high-energy phosphate group from ATP to each molecule. The two 3PG molecules are then converted to a RuBP molecule, which stays in the cycle for another around of CO_2, and an organic carbon unit [C]. Three rounds of Calvin cycle lead fixation of 3 CO_2 molecules and the production of a G3P molecule, which can be further converted to glucose, lipids and other cell materials.

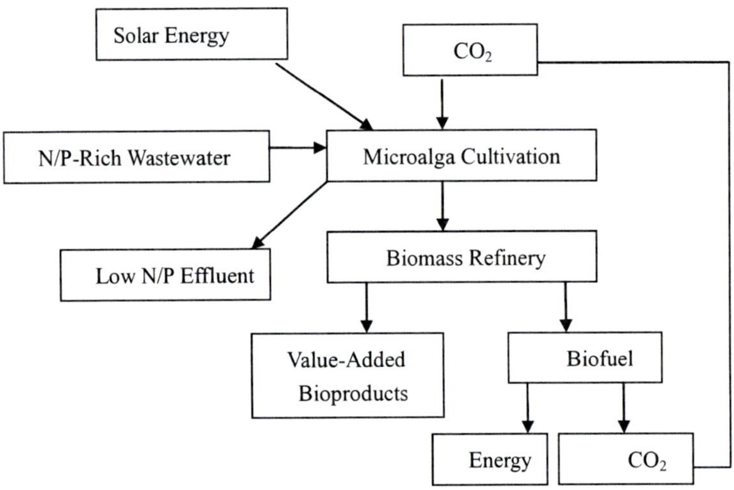

Figure 3.1. A conceptual microalgael system for combined biofuels production, CO_2 bio-mitigation and N/P removal from wastewater. (Wang B.et al.,2008) Inputs: carbon source, CO_2; nitrogen and phosphorus sources, N/P-rich wastewater; energy source, solar energy. Outputs: Low N/P effluent, value-added bioproducts, and biofuels.

3.3. SOURCES OF CO_2

Microalgae can fix carbon dioxide from different sources, which can be categorized as 1) CO_2 from the atmosphere; 2) CO_2 from industrial exhaust gases (e.g., flue gas and flaring gas); and 3) fixed CO_2 in the form of soluble carbonates (e.g., $NaHCO_3$ and Na_2CO_3).

Traditionally, microalgae are cultivated in closed systems or open ponds, which are aerated or exposed to air to allow microalgae to capture carbon dioxide from the atmosphere for cell growth. Since the atmosphere contains only 0.03 - 0.06% CO_2, it is expected that mass transfer limitation could slow down the cell growth of microalgae (Chelf, Brown et al. 1993). On the other hand, industrial exhaust gases such as flue gas contains up to 15% CO_2, providing a CO_2-rich source for microalgael cultivation and a potentially more efficient route for CO_2 bio-fixation. The third route is to fix CO_2 by chemical reaction to produce carbonates (e.g., Na_2CO_3) and use the latter as the carbon source for microalgael cultivation.

A number of microalgael species have been shown to be able to utilize carbonates such as Na_2CO_3 and $NaHCO_3$ for cell growth (Ginzburg 1993; Merrett, Nimer et al. 1996; Emma Huertas, Colman et al. 2000). Some of these species typically have high extracellular carboanhydrase activities (Emma Huertas, Colman et al. 2000), which is responsible for the conversion of carbonate to free CO_2 to facilitate CO_2 assimilation. In addition, the direct uptake of bicarbonate by an active transport system has been found in several species (Colman and Rotatore 1995; Merrett, Nimer et al. 1996). Adoption of Carbonate-Utilizing strains for CO_2 fixation could be advantageous in many aspects: 1) CO_2 released in nighttime from industrial facilities could be converted to carbonate salts and stored for conversion in daytime; 2) since only a limited number of microalgael species thrive in media containing high concentration of carbonate salts, species control (i.e., preventing wild type microalgael species from contaminating the cultivation system) is relatively simple; 3) most of these species have high pH optima (in the range of 9.0 to 11), further simplifies species control (Ginzburg 1993).

Flue gases from power plant are responsible for more than 7% of the total world CO_2 emissions (Sakai, Sakamoto et al. 1995). Carbon dioxide in flue gas is available at little or no cost. As estimated by the IPCC criteria, the CO_2 concentration of flue gas is up to 15% (Maeda, Owada et al. 1995). Therefore, it would be beneficial if microalgae are tolerant to elevated CO_2 level should they be used for CO_2 fixation from flue gases (Maeda, Owada et al. 1995).

An early review on flue gas tolerance by microalgae indicated that high levels of CO_2 were tolerated by many microalgael species and that moderate levels of SO_x and NO_x (up to 150ppm) were also well-tolerated (Matsumoto, Hamasaki et al. 1997). *Chlorococcum littorale,* a marine algae, showed exceptional tolerance to high CO_2 concentration of up to 40% (Murakami and Ikenouchi 1997; Iwasaki, Hu et al. 1998). Microalgae *Scenedesmus obliquus* and *Chlorella kessleri*, separated from the waste treatment ponds of the Presidente Médici coal fired thermoelectric power plant, also exhibited good tolerance to high CO_2 contents (de Morais and Costa 2007). *Chlorella kessleri* showed maximum specific growth rates (μ_{max}) of 0.267d^{-1} and biomass productivity of approximately 0.087 g L^{-1} d^{-1} when cultivated with 6% (V/V) and 12% (V/V) CO_2 and with a maximum biomass productivity of 0.085 g L^{-1} d^{-1} was achieved at 6% CO_2. These two microalgae also grew well when the culture was supplemented with enriched air stream contained up to 18% CO_2, indicating their great potentials for CO_2 fixation from CO_2-rich streams. It was also reported (de Morais and Costa 2007) that *Scenedesmus obliquus* and *Spirulina sp.* showed good capacities to fix carbon dioxide when they were cultivated at 30°C in a temperature-controlled three-stage serial tubular photo-bioreactor. For *Spirulina sp.*, the maximum specific growth rate and maximum productivity were 0.44 d^{-1} and 0.22 g L^{-1} d^{-1}, with both 6% (v/v) carbon dioxide and 12% (v/v) carbon dioxide, respectively, while the maximum cell concentration was 3.50 g dry cell L^{-1} with both CO_2 concentrations. For *S. obliquus*, the corresponding maximum growth rate and maximum productivity were 0.22 d^{-1} and 0.14 g L^{-1} d^{-1}, respectively. Murakami and Ikenouchi (1997) selected more than 10 strains of microalgae with high capability of fixing CO_2 by extensive screening. Two green algael strains, *Chlorella* sp. UK001 and *Chlorococcum littorale*, showed high CO_2 fixation rates exceeding 1 g CO_2 L^{-1} d^{-1}. *Botryococcus braunii* SI-30, which showed the ability of producing high content of hydrocarbons, was recommended as a promising candidate for combined CO_2 mitigation and biofuel production.

The tolerance of microalgae to relatively high temperature is very important in reducing cooling costs of the feeding flue gases released from industrial facilities at high temperature. These thermotolerant strains may also simplify species-control because the temperature optima of most microalgael species locate in the range of 20-30 °C. A few thermotolerant strains have been selected. For instance, several unicellular green algael strains, identified as species of *Chlorella*, were isolated from hot springs in Japan (Sakai, Sakamoto et al. 1995). These strains grew at temperatures up to 42 °C and in air containing more than 40% CO_2. Their tolerance to both high temperature and high CO_2 content makes them potentially the appropriate microbial cellar reactors for bio-CO_2 mitigation from flue gas.

Table 3.3 summarizes a few microalgael strains that have been studied for CO_2 bio-mitigation. Some of these strains can tolerate high temperature and high CO_2 in the gas stream. CO_2 fixed through photosynthesis are converted to different organic cell components including carbohydrates, lipids, proteins, and nucleic acids (Spolaore, Joannis-Cassan et al. 2006). Although the cell carbon content varies with microalgael strains, media and cultivation conditions, it changes in a relatively small range and the law of material conservation allows us to calculate CO_2 fixation rate from biomass productivity at given cell carbon content. In Table 3.3, such calculations were conducted using a reported biomass molecular formula, $CO_{0.48}H_{1.83}N_{0.11}P_{0.01}$ (Chisti 2007), when direct data on CO_2 fixation rate was not available, based on the assumption that CO_2 fixed in the form of extracellular products was negligible.

Table 3.3. Some microalgael strains studied for CO_2 bio-mitigation (Wang B. et al., 2008)

Microalgae	CO_2 %	T °C	P $gl^{-1}d^{-1}$	P_{CO2} $g\,l^{-1}d^{-1}$	Reference	Note
Chlorococcum littorale	40	30	N/A	1.0	(Murakami and Ikenouchi 1997; Iwasaki, Hu et al. 1998)	
Chlorella kessleri	18	30	0.087	0.163*	(de Morais and Costa 2007)	
Chlorella sp. UK001	15	35	N/A	>1	(Murakami and Ikenouchi 1997)	
Chlorella vulgaris	15		N/A	0.624	(Yun, Lee et al. 1997)	Artificial Wastewater
Chlorella vulgaris	air	25	0.040	0.075*	(Scragg, Illman et al. 2002)	Watanabe's medium
Chlorella vulgaris	air	25	0.024	0.045*	(Scragg, Illman et al. 2002)	Low-N medium
Chlorella sp.	40	42	N/A	1.0	(Sakai, Sakamoto et al. 1995)	
Dunaliella	3	27	0.17	0.313*	(Kishimoto, Okakura et al. 1994)	High salinity, ß-carotene
Haematococcus pluvialis	16-34	20	0.076	0.143	(Huntley and Redalje 2007)	Commercial scale, outdoor
Scenedesmus obliquus	air	-	0.009	0.016	(Go?mez-Villa, Voltolina et al. 2005)	Wastewater, outdoor, winter
Scenedesmus obliquus	air	-	0.016	0.031	(Go?mez-Villa, Voltolina et al. 2005)	Wastewater, outdoor, summer
Botryococcus braunii	-	25-30	1.1	>1.0	(Murakami and Ikenouchi 1997)	Accumulating Hydrocarbon
Scenedesmus obliquus	18	30	0.14	0.26	(de Morais and Costa 2007)	
† *Spirulina sp.*	12	30	0.22	0.413*	(de Morais and Costa 2007)	

* Calculated from the biomass productivity according to equation, CO_2 Fixation Rate (P_{co2}) = 1.88×Biomass Productivity (P), which is derived from the typical molecular formula of microalgael biomass, $CO_{0.48}H_{1.83}N_{0.11}P_{0.01}$ (Chisti 2007).

† All species except Spirulina sp., which is a prokaryotic cyanobacteria (Cyanophyceae) species, are eukaryotic green algae (Chlorophyta) species (NCBI website).
Not specified or not controlled

CO_2 mitigation by microalgae can be achieved either by microalgae in their natural habitats such as oceans and lakes or in microalgael farming facilities.

3.4. Ocean Fertilization

Oceans are the largest habitats of microalgae. In fact, the marine phytoplankton, which is comprised primarily of unicellular marine microalgae, is responsible for half of the planetary annual carbon fixation (approximately 5.0 x 10^{11} tons of carbon per annum). The oceans are in general rich in micronutrients. However, their biological productivity is limited by macronutrient such as iron. Ocean fertilization, which refers to the strategy of enriching ocean waters with limiting nutrients such as irons, is therefore proposed to enhance ocean biological production and CO_2 sequestering (Chaumont 1993; Zeebe and Archer 2005; Glibert, Azanza et al. 2008).

It has been well established that ocean fertilization can increase the primary production and hence CO_2 fixation over a significant fraction of the oceans (Liss, Chuck et al. 2005; Zeebe and Archer 2005)) and many advocates see ocean fertilization as modern society's last hope to slow global warming. However, this strategy is not free of controversies. The two major concerns are 1) the lack of large scale experimental data with respect to the efficiency of ocean fertilization (Aumont and Bopp, 2006) and 2) the uncertainty with regards to the biochemical and ecological side effects of ocean fertilization (Fuhrman and Capone, 1991, Liss et al., 2005, Glibert et al., 2008). It is suggested that ocean fertilization should be regulated to ensue precautions are taken in this practice (Orbach, 2008). Inconsistency of current international laws in this field is also recognized as a challenge (Freestone and Rayfuse, 2008)

3.5. Microalgael Biomass Harvesting and Drying

Due to light limitation, the biomass concentration of microalgae suspensions are usually low, in the range of 0.5 – 3.0 g/l. This low biomass concentration, in combination of the small cell size of microalgae, makes the biomass harvesting and drying costly and energy consuming. Different technologies, including chemical flocculation (Knuckey et al., 2006), biological flocculation (Divakaran and Pillai, 2002), filtration (Molina Grima et al., 2003), centrifugation (Olaizola, 2003), and ultrasonic aggregation (Bosma et al., 2003) have been investigated for microalgael biomass harvesting.

In general, chemical and biological flocculation require only low operating costs; however, they have the disadvantage of requiring long processing time and the risk of bioreactive product decomposition. On the other hand, filtration, centrifuge, and ultrasonic flocculation are more efficient but more costly. The selection of appropriate harvesting technology depends on the value of the target products, the biomass concentration, and the size of microalgael cells of interest.

Biomass drying before further extraction and/or thermochemical processing is another step that needs to be taken into consideration. Sun drying is probably the cheapest drying method that has been employed for the processing of microalgael biomass (Millamena et al., 1990, Prakash et al., 1997). However, this method takes long drying time, requires large drying surface, and risks the loss of some bioreactive products. Low pressure shelf drying is another low-cost drying technology that has been investigated (Prakash et al., 1997). It is nevertheless ineffective, requires long processing time and risk decomposition of bioproducts. More efficient but more costly drying technologies that have been investigated for drying microalgae include drum drying (Prakash et al., 1997), spray drying (Desmorieux and Decaen, 2005, Leach et al., 1998), fluidized bed drying (Leach et al., 1998), freezing drying (Millamena et al., 1990) and refractance window dehydration technology (Nindo and Tang, 2007). It is important to find the balance between the drying efficiency and cost-effectiveness.

3.6. Strategies to Enhance the Cost-Effectiveness of CO_2 Sequestration using Microalgael Farming

The microalgae-for-CO_2-sequestration strategy offers numerous advantages in comparison to biomitigation of CO_2 using agricultural or forestry crops or conventional chemical/physical CO_2 sequestration approaches. Photosynthetic microorganisms in general have much higher growth rates, photosynthetic efficiency and CO_2 fixation abilities, thanks to their vast surface to volume ratio and simple structure, compared to conventional forestry, agricultural, and aquatic plants (Chisti, 2007, Borowitzka, 1999). It has been reported that microalgae have the ability to fix CO_2 while capturing solar energy with an efficiency 10 to 50 times greater than that of terrestrial plants (Li et al., 2008b, Usui and Ikenouchi, 1997). It could completely recycle CO_2 (Fig. 3.1) since carbon dioxide is converted into the chemical energy via photosynthesis, which can be converted to fuels using existing technologies (Demirbas, 2004). In comparison, the chemical reaction-based CO_2 mitigation approaches, as discussed above, have disposal problems because both the captured CO_2 and the wasted absorbents need to be disposed of (Yeh et al., 2001, Bonenfant et al., 2003). Chemical reaction-based CO_2 mitigation approaches are energy consuming and costly processes (Resnik et al., 2004, Lin et al., 2003) and the only economical incentive for CO_2 sequestrating using chemical reaction-based approach is the CO_2 credits to be generated under the Kyoto protocol. Very encouraging, CO_2 bio-mitigation via microalgael farming could potentially be made profitable from the production of biofuels and other novel bioproducts. Nevertheless, the large costs associated with microalgael farming are still the most significant obstacle toward the commercial implementation of this strategy and a few strategies could be employed to enhance its cost-effectiveness.

3.6.1. Combined CO_2 Bio-Mitigation with Biofuel Production

The merit of CO_2 bio-mitigation by microalgael farming locates primarily in the fact that biomass produced in the process of CO_2 fixation can be converted efficiently into biofuels for energy production (Li et al., 2008a). An estimate made in 2003 indicates that the costs of biofuel production are in general about 2.3 times more expensive than fossil fuels (Kondili and Kaldellis 2007). There is no doubt that fast technology development and the soaring energy prices have improved and will improve the situation rapidly and biofuel production from microalgae is deemed to be the most promising biofuel production strategy (Li Y. 2008). There is no doubt that global efforts from both the public and private sectors will be continued and accelerated in order to make biofuels from microalgae a practical replacement of fossil fuels in the near future.

3.6.2. Biorefinery: The High-Value Co-Product Strategy

The term biorefinery was coined to describe the production of a wide range of chemicals and biofuels from biomasses by the integration of bioprocessing and appropriate low environmental impact chemical technologies in a cost-effective and environmentally sustainable manner (Chisti, 2007). Examples include the two-phase conversion reaction of fructose to 5-hydroxymethylfurfural (Kamm, 2007), fermentative production of ethanol from sugars derived from cellulose and semi-cellulose (Qu, 2007), bio-oils and/or bio-syngas by the pyrolysis/gasification of woods or other types of biomasses (Mohan et al., 2006).

Microalgae have the capacity of producing a vast array of high-value bioactive compounds that can be used as pharmaceutical compounds, health foods, and natural pigments (Lee and Ohkita, 2003, Jiang, 2000). Some well studied examples include acetylic acids, β-carotene (Huang et al., 2006, Del Campo et al., 2007), vitamin B (Yue and Chen, 2005), ketocarotenoid astaxanthin (Kaewpintong et al., 2007), polyunsaturated fatty acids (Wen and Chen, 2003, Jiang et al., 2004), and lutein (Shi et al., 2002, Blanco et al., 2007) (see Table 1). The economical feasibility of microalgael biofuel production could therefore be enhanced by a high-value co-product strategy, which will involve sequentially the cultivation of microalgae in a microalgael farming facility (CO_2 mitigation), extracting bioreactive products from harvested algael biomass, thermal processing (pyrolysis or gasification), extracting high-value chemicals from the resulting liquid, vapour and/or solid phases, and reforming/upgrading biofuels for different applications. The employment of a high-value co-product strategy through the integrated biorefinery approach is expected to significantly enhance the overall cost-effectiveness of microalgael biofuel production.

3.6.3. Combination of Microalgael Cultivation with Wastewater Treatment

Combination of wastewater treatment and CO_2-fixation from microalgael biomass provides additional economic incentives due to the savings from chemicals (the nutrients) and the environment benefits (Mallick, 2002), which include: 1) microalgae have been shown to be efficient in nitrogen and phosphorus removal (Mallick, 2002) as well as in metal ion depletion, and combination of microalgae cultivation with wastewater treatment will significantly enhance the environmental benefit of this strategy; and 2) it will lead to savings in term of minimizing the use of chemicals such as sodium nitrate and potassium phosphorus as exogenous nutrients and 3) it will result in savings of the precious freshwater resources. Figure 3.1 depicts a conceptual flow chart for the complete "recycling" of CO_2 for solar energy capturing. Finally, enhance the primary production of oceans by means of ocean fertilization may provide a cost effective strategy of large scale CO_2 sequestrating that is efficient enough to cope with the global warming. The potential of wastewater treatment by microalgae have been investigated by a few researchers (Oswald, 1973, Benemann et al.,

1977) with several microalgael strains tested for this purpose. For instance, *B. braunii* was shown to be able to remove nitrogen and phosphorus from secondarily treated wastewater (which has been eliminated the easily settled materials and organic materials) in a batch system and a continuous bioreactor system with hydrocarbon production. *Chlorella vulgaris* (Yun et al., 1997) was cultivated in wastewater discharged from a steel-making plant with the aim of developing an economically feasible system to remove ammonia from wastewater and CO_2 from flue gas simultaneously. CO_2 fixation and ammonia removal rates were estimated as 26.0 g CO_2 m^{-3} h^{-1} (0.624 g l^{-1} day^{-1}) and 0.92 g NH_3 m^{-3} h^{-1}, respectively, when the algae was cultivated in wastewater supplemented with 46.0 g PO_4^{-3} m^{-3} without pH control at 15% (v/v) CO_2.

Microalgael biomass production by outdoor cultivation of *Scenedesmus obliquus* in artificial wastewater under the winter and summer tropical conditions of Mazatlán, Sinaloa, Mexico has been reported (Gomez-Villa et al., 2005). The biomass concentrations were 26 an 43 mg l^{-1} after three days of cultivation in the winter and summer, respectively, corresponding to biomass productivities of 9 and 16 mg l^{-1} day^{-1}, which were equivalent to CO_2 fixation rates of 16.07 and 31.0 mg CO_2 l^{-1} day^{-1} when the aforementioned typical biomass molecular formula was adopted for the calculation. The final dissolved nitrogen concentrations were 53% of the initial value in winter and 21 % in summer. Phosphorus was removed only during the day, with a total abatement of 45% in winter and 73% in summer.

3.7. Future Trends

CO_2 fixation using fast growing autotrophic microorganisms, especially microalgael species, in their natural habitats or in artificial cultivation systems provides a very promising alternative for mitigation of CO_2, the most prominent greenhouse gas. The primary merit of this strategy lays in the fact that via the cultivation of microalgae, CO_2 mitigation and biofuel production could be combined in an economically feasible and environmentally sustainable manner. The feasibility of this strategy could be further enhanced by fixing CO_2 from industrial exhaust gases such as flue gases and by integrating microalgael cultivation with wastewater treatment. Since bio-fixation CO_2 by microorganisms will play an important role in preventing and curing CO_2 emission and global warming. Employment of the high-value co-product strategy is also expected to enhance the economic viability of the CO_2 biofixation strategy.

Nevertheless, there are some limitations in the microalgael CO_2 mitigation strategy that needs to be addressed. The primarily concern associated with microalgael farming is the relatively high costs of microalgael cultivation, biomass harvesting, drying, and downstream processing. Extensive studies are expected to generate sufficient technology advances for cost-effective CO_2 fixation using microalgae.

REFERENCES

Blauwhoff, P. M. M., G. F. Versteeg, et al. (1984). "A study on the reaction between CO_2 and alkanolamines in aqueous solutions." *Chem. Eng. Sci.* 39(2): 207-225.

Bonenfant, D., M. Mimeault, et al. (2003). "Determination of the structural features of distinct amines important for the absorption of CO2 and regeneration in aqueous solution." *Industrial and Engineering Chemistry Research* 42(14): 3179-3184.

Borowitzka, M. A. (1999). "Commercial production of microalgae: ponds, tanks, tubes and fermenters." *Journal of Biotechnology* 70(1-3): 313-321.

Chelf, P., L. M. Brown, et al. (1993). "Aquatic biomass resources and carbon dioxide trapping." *Biomass & bioenergy* 4(3): 175-183.

Chisti, Y. (2007). "Biodiesel from microalgae." *Biotechnology Advances* 25(3): 294-306.

Colman, B. and C. Rotatore (1995). "Photosynthetic inorganic carbon uptake and accumulation in two marine diatoms." *Plant, Cell and Environment* 18(8): 919-924.

Cooper, C. D. and F. C. Alley (1994). *Air Pollution Control: A Design Approach.*

de Morais, M. G. and J. A. V. Costa (2007). "Biofixation of carbon dioxide by Spirulina sp. and Scenedesmus obliquus cultivated in a three-stage serial tubular photobioreactor." *Journal of Biotechnology* 129(3): 439-445.

de Morais, M. G. and J. A. V. Costa (2007). "Isolation and selection of microalgae from coal fired thermoelectric power plant for biofixation of carbon dioxide." *Energy Conversion and Management* 48(7): 2169-2173.

Demirbas, A. (2004). "Current technologies for the thermo-conversion of biomass into fuels and chemicals." *Energy Sources* 26(8): 715-730.

Emma Huertas, I., B. Colman, et al. (2000). "Active transport of CO_2 by three species of marine microalgae." *Journal of Phycology* 36(2): 314-320.

Ginzburg, B. Z. (1993). "Liquid fuel (oil) from halophilic algae: a renewable source of non- polluting energy." *Renewable Energy* 3(2-3): 249-252.

Go?mez-Villa, H., D. Voltolina, et al. (2005). "Biomass production and nutrient budget in outdoor cultures of Scenedesmus obliquus (Chlorophyceae) in artificial wastewater, under the winter and summer conditions of Mazatlan, Sinaloa, Mexico." *Vie et Milieu* 55(2): 121-126.

Gupta, H. and L. S. Fan (2002). "Carbonation-calcination cycle using high reactivity calcium oxide for carbon dioxide separation from flue gas." *Industrial and Engineering Chemistry Research* 41(16): 4035-4042.

Huntley, M. E. and D. G. Redalje (2007). "CO_2 mitigation and renewable oil from photosynthetic microbes: A new appraisal." *Mitigation and Adaptation Strategies for Global Change* 12(4): 573-608.

Iwasaki, I., Q. Hu, et al. (1998). "Effect of extremely high-CO_2 stress on energy distribution between photosystem I and photosystem II in a 'high-CO2' tolerant green algae, Chlorococcum littorale and the intolerant green algae Stichococcus bacillaris." *Journal of Photochemistry and Photobiology B: Biology* 44(3): 184-190.

Kishimoto, M., T. Okakura, et al. (1994). "CO_2 fixation and oil production using micro-algae." *Journal of Fermentation and Bioengineering* 78(6): 479-482.

Kondili, E. M. and J. K. Kaldellis (2007). "Biofuel implementation in East Europe: Current status and future prospects." *Renewable and Sustainable Energy Reviews* 11(9): 2137-2151.

Krumdieck, S., J. Wallace, et al. (2008). "Compact, low energy CO_2 management using amine solution in a packed bubble column." *Chemical Engineering Journal* 135(1-2): 3-9.

Li Y., H. M., Wu N, Lan CQ, Dubois-Calero N. (2008). "Biofuels from Microalgae." *Biotech Prog.* in press.

Lin, C. C., W. T. Liu, et al. (2003). "Removal of carbon dioxide by absorption in a rotating packed bed." *Industrial and Engineering Chemistry Research* 42(11): 2381-2386.

Maeda, K., M. Owada, et al. (1995). "CO_2 fixation from the flue gas on coal-fired thermal power plant by microalgae." *Energy Conversion and Management* 36(6-9): 717-720.

Mallick, N. (2002). "Biotechnological potential of immobilized algae for wastewater N, P and metal removal: A review." *BioMetals* 15(4): 377-390.

Matsumoto, H., A. Hamasaki, et al. (1997). "Influence of CO_2, SO_2 and no in flue gas on microalgae productivity." *Journal of Chemical Engineering of Japan* 30(4): 620-624.

Merrett, M. J., N. A. Nimer, et al. (1996). "The utilization of bicarbonate ions by the marine microalgae Nannochloropsis oculata (Droop) Hibberd." *Plant, Cell and Environment* 19(4): 478-484.

Murakami, M. and M. Ikenouchi (1997). "The biological CO_2 fixation and utilization project by RITE (2): Screening and breeding of microalgae with high capability in fixing CO_2." *Energy Conversion and Management* 38(SUPPL. 1): S493-S497.

Ragauskas, A. J., C. K. Williams, et al. (2006). "The path forward for biofuels and biomaterials." *Science* 311(5760): 484-489.

Resnik, K. P., J. T. Yeh, et al. (2004). "Aqua ammonia process for simultaneous removal of CO_2, SO_2 and NOx." *International Journal of Environmental Technology and Management* 4(1-2): 89-104.

Sakai, N., Y. Sakamoto, et al. (1995). "Chlorella strains from hot springs tolerant to high temperature and high CO_2." *Energy Conversion and Management* 36(6-9): 693-696.

Scragg, A. H., A. M. Illman, et al. (2002). "Growth of microalgae with increased calorific values in a tubular bioreactor." *Biomass and Bioenergy* 23(1): 67-73.

Seefeldt, L. C. (2007). "Utah group plans to make biodiesel from algae." *Industrial Bioprocessing* 29(3): 5-6.

Shi, M. and Y. M. Shen (2003). "Recent progresses on the fixation of carbon dioxide." *Current Organic Chemistry* 7(8): 737-745.

Skja?nes, K., P. Lindblad, et al. (2007). "BioCO_2 - A multidisciplinary, biological approach using solar energy to capture CO_2 while producing H2 and high value products." *Biomolecular Engineering* 24(4): 405-413.

Spolaore, P., C. Joannis-Cassan, et al. (2006). "Commercial applications of microalgae." *Journal of Bioscience and Bioengineering* 101(2): 87-96.

Usui, N. and M. Ikenouchi (1997). "The biological CO_2 fixation and utilization project by RITE(1): Highly-effective photobioreactor system." *Energy Conversion and Management* 38(SUPPL. 1): S487-S492.

Wang, B., Y. Li, et al. (2008). "CO_2 bio-mitigation using microalgae." *Applied Microbiology and Biotechnology* 79(5): 707-718.

Yeh, J. T., H. W. Pennline, et al. (2001). "Study of CO_2 absorption and desorption in a packed column." *Energy and Fuels* 15(2): 274-278.

Yun, Y. S., S. B. Lee, et al. (1997). "Carbon dioxide fixation by algael cultivation using wastewater nutrients." *Journal of Chemical Technology and Biotechnology* 69(4): 451-455.

Chapter 4

BIOFUELS FROM MICROALGAE

Most the energy carriers on earth originate ultimately from the sun. Examples of these sun-origin energies include fossil fuels, solar energy, biofuels, wind power, and hydropower. Hypothetically, fossil fuels such as coal, oil and natural gas were converted from biomass under high pressure high temperature condition at the absence of oxygen when large quantities of biomass were buried beneath the earth surface due to catastrophic changes of the earth shell and were accumulated over billions of years in the past of earth history and are therefore practically unrenewable. The other sun-origin energies are renewables and shall contribute the solution of the sustainable energy cycle in the future. The wind energy and hydropower are available at limited scales and solar energy, when converted to electricity directly using semi-conductor batteries, is ultimately limited by the availability of silicon, a rare element on earth. As a result, biofuels derived from different biomasses have been regarded as the most promising renewable energy sources.

A large diversity of different biofuels has been developed in the past few decades and some of them have achieved great commercial success. Most important examples of these biofuels include bio-ethanol/butanol, biodiesel, biohydrogen, bio-gas, bio-oil, and bio-syngas. Biomasses from different sources, including forestry, agricultural, and aquatic sources have been investigated as the feedstock for the production of different biofuels. However, burning fuels derived from existing biomass has an environmental impact similar to the combustion of fossil fuels in terms of the carbon cycle, i.e., conversion of fixed carbon into CO2. In addition, depletion of certain existing biomasses (e.g., wood) without appropriate compensation (e.g., replanting) may result in massive biomass deficit, creating serious environmental problems (e.g., deforestation).

Conventional terrestrial plants are not very efficient in capturing solar energy. It was estimated that switchgrass, the fastest-growing terrestrial crops, can convert solar energy to biomass-energy at a yearly rate of no more than 1 W/m2, less than 0.5% of the solar energy received at a typical mid-latitude region (200–300 W/m2) (UN 2003; Lewis and Nocera 2006). On the other hand, studies have shown the photosynthetic efficiency of microalge could well be in the range of 10 - 20% or higher (Richmond 2000),(Huntley and Redalje 2007). Furthermore, recent studies showed that the extra N2O entering the atmosphere as a result of using nitrogen fertilizers to produce crops for biofuels, when calculated in "CO2-equivalent" global warming terms, and compared with the quasi-cooling effect of "saving" emissions of fossil fuel derived CO2, could contribute as much or more to global warming by N2O emissions than cooling by fossil fuel savings (Crutzen, Mosier et al. 2007). These concerns may be addressed by using fast-growing microalgael species for biofuel production.

Microalgae have high growth rates and photosynthetic efficiencies due to their simple structures. It is estimated that the biomass productivity of microalgae can be 50 times more than that of switch grass (Demirbas 2006; Nakamura 2006). Biofuel production using microalgael farming offers the following advantages: 1) high growth rate of microalgae makes it possible to satisfy the massive demand on biofuels using limited land resources without causing potential biomass deficit; 2) microalgael cultivation consumes less water than land crops; 3) tolerance of microalgae to high-CO2 content in gas streams allows high-efficiency CO2 mitigation; 4) nitrous oxide release could be minimized when microalgae are used for biofuel production; 5) microalgael farming could be potentially more cost effective than conventional farming. On the other hand, one of the major disadvantages of microalgae for biofuel production is the low biomass concentration in the algael culture due to the limit of light penetration, which, in combination with the small size of algael cells, makes the harvest of algael biomasses relatively costly. Large water content of harvested algael biomass also means its drying would be an energy-consuming process. The significantly high capital costs of and the rather intensive cares required by a microalgael farming facility than a conventional agricultural farm would likely to invoke higher investments and operating costs. Nevertheless, these problems are expected to be overcome or minimized by technology development.

Microalgae was initially examined as a potential replacement fuel source for fossil fuels in the 1970's amidst the gas scare (W. Barkley 1987), but prohibitive production costs and limitations discouraged the commercial development of algae-based fuel production. Subsequent studies, continued through the 1980's and heightened in the last 15 years, illustrate that research developments are enabling the commercial potential of microalgae to shift from aquaculture, fine chemicals, and health food (De la Noue 1988) to fuel production. In the following sections, we discuss different biofuels and the ways to produce these biofuels from microalgael feedstock.

4.1. BIODIESEL

Biodiesel is produced by a mono-alcoholic transesterification process, in which triglycerides reacts with a mono-alcohol (most commonly methanol or ethanol) with the help of basic, acidic and enzymatic catalysts (Meher, Vidya Sagar et al. 2006; Demirbas 2007). It has similar combustion properties to diesel (Marchetti, Miguel et al. 2007) and has been produced commercially or in backyard facilities to fuel vehicles. Significant technical advances have been achieved to optimize the trans-esterification process. For instance, Canadian researchers in the Department of Chemical Engineering at the University of Ottawa have developed a novel two-phase membrane reactor (Dube, Tremblay et al. 2007), which exploits the immiscibility of canola oil in methanol to enable the separation of reaction products (biodiesel/glycerol) from the residual canola oil. The two-phase membrane reactor was particularly useful in removing unreacted canola oil from the product, yielding high purity biodiesel and shifting the reaction equilibrium to the product side. Nevertheless, one major challenge of biodiesel production is the high costs of feedstock. Currently, biodiesel production relies on animal fats and plant oils. However, replacing all the transport fuel consumed in the United States with biodiesel will require 0.53 billion m3 of biodiesel annually at the current rate of consumption. Oil crops, waste cooking oil and animal fat cannot realistically satisfy this demand. This is demonstrated in Table 4. Using the average oil yield per hectare from various crops, the cropping area needed to meet 50% of the U.S. transport fuel needs is calculated in column 3 (Table 5). In column 4 of Table 5, this area is expressed as a percentage of the total cropping area of the United States. If oil palm, a high-yielding oil crop can be grown, 24% of the total cropland will need to be devoted to its cultivation to meet only 50% of the transport fuel needs. Clearly, oil crops cannot replace petroleum derived liquid fuels in the foreseeable future. This scenario changes dramatically, if microalgae are used to produce biodiesel. Between 1 and 3% of the total U.S. cropping area would be sufficient for producing algael biomass that satisfies 50% of the transport fuel needs. The microalgael oil yields given in Table 5 are based on experimentally demonstrated biomass productivity in photobioreactors, as discussed later in this article. Actual biodiesel yield per hectare is about 80% of the yield of the parent crop oil given in Table 4.1.

In view of Table 4.1, microalgae appear to be the only source of biodiesel that has the potential to completely displace fossil biodiesel. Unlike other oil crops, microalgae grow extremely rapidly and many are exceedingly rich in oil. Microalgae commonly double their biomass within 24h. Biomass doubling times during exponential growth are commonly as short as 3.5h. Oil content in microalgae can exceed 80% by weight of dry biomass. Oil levels of 20-50% are quite common (Table 4.2). Oil productivity, that is the mass of oil produced per unit volume of the microalgael broth per day, depends on the algael growth rate and the oil content of the biomass. Microalgae with high oil productivities are desired for producing biodiesel.

Table 4.1. Comparison of some sources of biodiesel (Chisti, 2007)

Crop	Oil yield (L/ha)	Land area Needed (M ha)[a]	Present of existing US cropping area[a]
Corn	172	1540	846
Soybean	446	594	326
Canola	1190	223	122
Jatropha	1892	140	77
Coconut	2689	99	54
Oil palm	5950	45	24
Microalgae[b]	136,900	2	1.1
Microalgae[a]	58,700	4.5	2.5

a For meeting 50% of all transport fuel needs of the United States.
b 70% oil (by wt) in biomass.
c 30% oil(by wt) in biomass.

This agricultural approach will eventually compete for land resource against food industry. For instance, it was estimated that to produce 5.54 Mtoe (million tons of oil equivalent) of biodiesel (1.72% of the 321 Mtoe estimated EU-25 consumption for transportation fuels in 2003) would require 9.3 Mha of land for Canola (rapeseed) and sunflower cultivation. This is equivalent to 150% of the current land used for these crops in EU-25 (Kondili and Kaldellis 2007). On the other hand, some microalgael species could accumulate lipids to a significant portion of their biomass (30 – 50% on dry weight basis), serving as a promising alternative source of lipids for biodiesel production (Tornabene, Holzer et al. 1983; Miao and Wu 2006; Xu, Miao et al. 2006). It was estimated by Sheehan and his co-workers that microalgael farming using 200,000 hectares of land would allow the production of a quad (i.e., a quadrillion BTU) of fuel in the form of biodiesel (J Sheehan, 1998). To put this in perspective, one quad is approximately 1/8 of Canada's total energy consumption in 2004 (8543.3×1015 J (Canada 2006)).

Table 4.2. Oil content of some microalgae (Chisti, 2007)

Microalgae	Oil content (% dry wt)
Botrycococcus braunii	25-75
Chlorella sp.	28-32
Crypthecodinium cohnii	20
Cylindrotheca sp.	16-37
Dunaliella primolecta	23
Isochrysis sp.	25-33
Monallanthus salina	>20
Nannochloris sp.	20-35
Nannochloropsis sp.	31-68
Neochloris oleoabundans	35-54
Nitzschia sp.	45-47
Phaeodactylum tricornutum	20-30
Schizochytrium sp.	50-77
Tetraselmis sueica	15-23

Christi discussed the economics and quality constraints of biodiesel from microalgae in his recent review paper (Chisti 2007). He pointed out that the cost of growing microalgae for biofuel production must be drastically reduced to compete directly with traditional sources. It is essential to consider the other roles algael cultures can play concurrently with biofuel production and the long term benefits this entails. It is interesting to notice that, even though the two major project sponsored by the US government and the Japanese government concluded that algael oil was not economically feasible (Huntley and Redalje 2007), the private sector has moved forward in building commercial facilities to produce biodiesel using algael oils. It was reported that a privately funded US$20 million program has engineered, built, and successfully operated for several years a commercial-scale (2 ha), modular, production system coupling photobioreactors with open ponds in a two-stage process to produce Haematococcus pluvialis for biodiesel production with a annual averaged rate of achieved microbial oil production equivalent to 420 GJ ha -1 yr-1, which exceeds the most optimistic estimates of biofuel production from plantations of terrestrial "energy crops." The maximum production rate achieved to was equivalent to 1014 GJ ha-1 yr-1. It was claimed that a rate of 3200 GJ ha-1 yr-1 is feasible using species with known performance characteristics under conditions that prevail in the existing production system, a rate possible to replace reliance on current fossil fuel usage equivalent to ~ 300 EJ yr -1 and eliminate fossil fuel emissions of CO_2 of ~ 6.5 Gigatons of Carbon (GtC) per year using only 7.3% of the surplus arable land projected to be available by 2050 (Huntley and Redalje 2007). Some algael biodiesel processes, such as the one being developed at the University of Utah, are expected to be cost-competitive with regular diesel by 2009 (Seefeldt 2007). Wu (2007) in China reported the production of biodiesel using chlorella protothecoides at a scale of 11,000 L. However, they adopted a heterotrophic cultivation strategy, which does not necessarily fulfill the mandate of microalgael farming to convert solar energy to biofuel.

Oils suitable as the feedstock of biodiesel production are triglycerides, which consist of three fatty acid residuals and a glycerol residue. Transestrification, which produces fatty acid methyl esters (FAME, biodiesel) and glycerol, requires 3 mol of alcohol for each mole of triglyceride to produce 1 mol of glycerol and 3 mol of FAME. Industrial processes commonly use 6 mol of methanol for each mole of triglyceride to ensure that the reaction is driven in the direction of FAME formation and high conversion of triglycerides. The yield of FAME on the basis of triglycerides typically exceeds 98% (w/w).

Transesterification can be catalyzed by acids, alkalis or lipase enzymes. Alkali-catalyzed transesterification is about 4000 times faster than the acid catalyzed reaction and are commonly used as commercial catalysts. Nevertheless, alkali can form soup with free fatty acids, which exist in most commercial oils, especial in waste oils. Existence of soup drastically increases difficulties in downstream processing. A common technique of deal with this problem is to use acid catalyze the esterification between free fatty acids and methanol and then use alkali to catalyze the transesterification reaction between triglycerides and methanol. It has been demonstrated that alkoxides such as sodium methoxide are even better catalysts than sodium hydroxide and are being increasingly used. Use of biocatalysts such as lipases offers important advantages, however, is not currently feasible because of the relatively high cost of the catalyst. Nevertheless, progress has been made in maintaining the activity of lipase in heterogeneous transesterification for prolonged enzymatic reaction, which led to drastic reduction of production costs. It is expected bio-transesterification will probably gain popularity in the near future.

4. 2. Bio-Hydrogen

Hydrogen, a carbon-free renewable energy carrier undergoes combustion without the release of green house gasses, is perceived to be one of the major alternatives to mitigating the current global environmental challenges. Currently, hydrogen finds applications primarily as raw material in hydrogenation processes for upgrading heavy oils, bitumen, oil sands, as propulsion fuel for space rocket and as a high grade feedstock for fuel cells. It also finds application in the manufacture of petrochemical products, as oxygen scavenger in reducing trace O_2 to prevent corrosion in metallurgy industry, and in food processing, semiconductor manufacturing and cooling of heavy duty industrial generators. Statistics show that the demand for hydrogen for these applications has been increasing consistently for the last five years, due mainly to stricter fuel quality standards, increased processing of heavy oil, and the advocacy for a hydrogen economy. It is foreseen that this trend will continue over the next decade (Das and Veziroglu 2008).

Hydrogen is the most abundant element in the universe and is found abundantly on earth in primarily in the form of organic compounds and water. In order to harness hydrogen from these diverse sources, a number of routes are being taken with each having its advantages and limitations. At present hydrogen is produced mainly from the reformation of natural gas via the water-gas shift reaction, which constitutes about 90%, while a limited amount is produced via other thermo-chemical processes and electrolysis of water. These pathways, however, are energy intensive, making them to be costly and with inherently high environmental impact compared to biological hydrogen production. Biohydrogen production, a renewable method of hydrogen production, is potentially cost-effective, sustainable, and of little or no environmental footprint. Often the main consideration for process selection is cost, operational flexibility, process efficiency, safety and environmental concerns. Table 4.3 summarizes these different routes and the availability of the different raw materials.

There are four different pathways for biological production of hydrogen (Meher Kotay and Das 2008): 1) Biophotolysis of water using microalgae, including eukaryotic microalgae and cyanobacteria; 2) Photodecomposition of organic compounds by photosynthetic bacteria; 3) Fermentative hydrogen production from organic compounds; and 4) Hybrid systems using photosynthetic and fermentative bacteria to produce hydrogen using a two-stage process. Figure 4.1 shows the hydrogen production pathway of cyanobacteria, which involves nitrogenase catalyzed hydrogen formation and hydrogenase catalyzed hydrogen uptake (Das and Vezirolu 2001).

Table 4.3. Hydrogen production routes, process, main feedstock and associated environmental impact (Sigfusson 2007)

Primary method	Process	Primary feed-stock	Other Feed-stock	energy	Quantity on earth (gigaton level)	Carbon footprint Or emissions
Thermal	Steam reforming	natural – gas, oil	Water	Heat	345 USGS 420	Carbon sequestration needed
	Gasification	coal	Water, oxygen	Steam at high T and P	6200	Carbon sequestration needed
	Pyrolysis	biomass		Moderately high temp steam		Essentially renewable
	Thermo-chemical	water		Nuclear	1,400 000 000 (USGS incl. oceans)	Nuclear waste disposal problems
Electro Chemical	Electrolysis	water		Renewables incl. solar wind, hydroelectric	1,400 000 000 (USGS incl. oceans)	emissions mostly related to life cycle
Bio logical	Photoelectron-Chemical	water	Algae strains	Direct sunlight	1,400 000 000 (USGS incl. oceans)	Minor emissions
	Photo-biological	water		Direct sunlight	1,400 000 000 (USGS incl. oceans)	No emissions
	Anaerobic digestion	biomass		High temp steam		LCA related minor
	Fermentative microorganisms	biomass		High temp steam		LCA related minor

Microalgae, like higher plants, fix carbon dioxide to produce oxygen in the presence of light to produce glucose and biomass via in the process termed photosynthesis, which involves two photosynthetic centres: photosystem II (PS II) and photosystem I (PS I). Some species can utilize the same photosynthesis system for the generation of hydrogen gas rather than for carbon fixation. In the process of hydrogen production by the pathway of water photo-splitting, water splitting and O_2 evolving system is carried out by PSII and the reducing power for the reduction of proton to molecular hydrogen is generated by PSI (Das and Vezirolu 2001). Since it was discovered by Garon et al. in early 1940s (Gaffron and Rubin 1942) that the eukaryotic unicellular green algae, *Scenedesmus obliquus*, was able to evolve molecular hydrogen by means of a hydrogenase in the light under anaerobic conditions, a large number of different microorganisms including green algae, cyanobacteria, photosynthetic bacteria and fermentative bacteria have been found to have the capacity to produce hydrogen through different pathways.

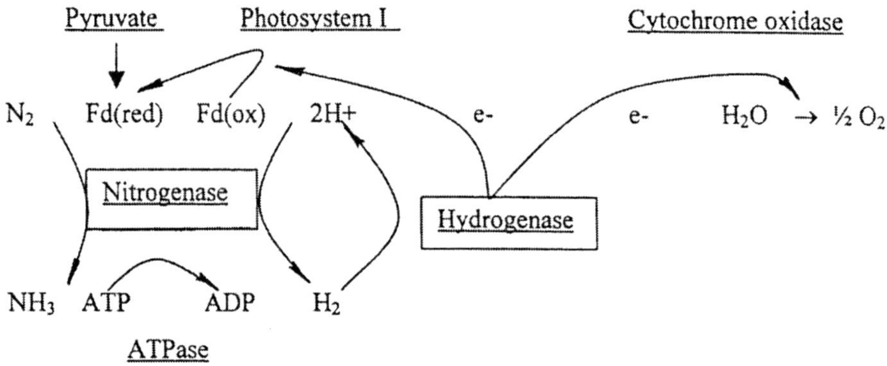

Figure 4.1. Nitrogenase-catalyzed hydrogen formation and hydrogenase-catalyzed H_2 uptake of cyanobacteria (Das and Vezirolu 2001)

4.3. Bioethanol

Bioethanol, and to a less extent biobutanol, has been being promoted as clean and renewable fuel that will reduce global warming and air pollution for decades. It is indeed the biofuel that achieved the most commercial success at present, which Brazil as the most successful model country. As shown in Table 4.4, the bioethanol production was approximately 13.1 billion gallons worldwide in 2007 and a nearly exponential growth of ethanol production was experienced in the U.S. in the last decade (Figure 4.2).

Production of ethanol via fermentation of sugars derived from corn, sweet potato, and sugar cane (e.g., molasses) is a proven technology and is currently the only means of commercial production of ethanol. This approach, however, competes for land and water resources with the food industry and is therefore not a sustainable means of biofuel production from a long term perspective.

Table 4.4. 2007 World Fuel Ethanol Production in Millions of Gallons

U.S.A	6498.6
Brazil	5019.2
European Union	570.3
China	486.0
Canada	211.3
Thailand	79.2
Colombia	74.9
India	52.8
Central America	39.6
Australia	26.4
Turkey	15.8
Pakistan	9.2
Peru	7.9
Argentina	5.2
Paraguay	4.7
Total	13101.7

Source: RFA, Changing the climate- ethanol industry outlook 2008

Extensive studies have been carried out on the utilization of liogcellusic biomass such as trees, grasses, corn stover, and wheat straw for bio-ethanol production. The lignocellulosic matrix is complex and recalcitrant to conversion but research in biorefining is advancing rapidly and commercial facilities are expected in the near future. There are two different pathways for the ethanol production from lignocellulosic biomass: 1) the biochemical process, which employs hydrolytic mechanisms to break apart the structural polysaccharides of the biomass to produce monosaccharides (i.e., saccharification) and then ferment the sugars using wildtype or engineered bacteria or yeasts to produce ethanol; and 2) the hybrid thermochemical/biochemical process, which involves thermochemical procedures to gasify biomass to produce bio-syngas and then use a special group of bacteria to convert syngas to ethanol. The largest challenge in utilizing lignocellulosic biomass for ethanol production is the high costs and energy consumption associated with the processes, which are expected to be overcome by technology development in the future.

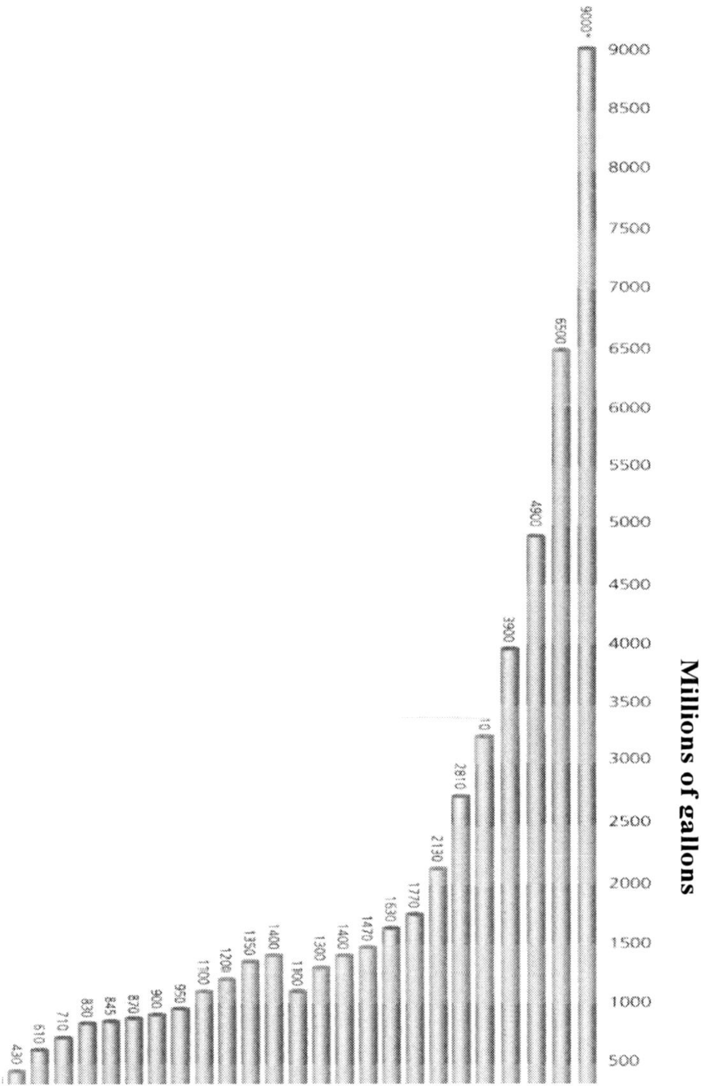

Figure 4.2. Fuel ethanol production in the U.S (source: Renewable Fuels Association, January 2009).

Cellulosic ethanol is made from the woody or structural parts of plants. Examples of these materials include agricultural residues, such as corn stover, cereal straws, and sugarcane bagasse; industrial waste, such as sawdust and paper pulp; forestry residues, such as small trees and excess wood; and energy crops, such as switchgrass, hybrid poplars, and willows, specifically grown for fuel production. The cell walls of these materials are comprised of long sugar chains (carbohydrates). The sugar in these cell walls is extracted through the introduction of enzymes or acids, and then converted to ethanol using microorganisms. Due to the complex structure of the cell walls, it is more difficult to break cellulosic materials into sugar, making them more expensive to convert to ethanol. Even with its complexity, cellulosic ethanol holds the promise of reducing petroleum use and lifecycle greenhouse gases (GHGs) emissions, and because it is made from non-food crops, cellulosic ethanol should not affect the food supply and should likely reduce impact on land and water. In 2007, The Unites States Department of Energy announced a multi-million dollar investment to build several cellulosic ethanol biorefineries to advance the commercial production of this and other biofuels (Renewable Fuels association, 2008). In Canada, Iogen Corp is building Canada's first cellulosic ethanol refinery, a $500-million plant in northern Saskatchewan that will use wheat straw to produce biofuel, electricity, enzymes and fertilizer (McCarthy S, 2008).

On April 25, 2008, Coskata Inc. announced that it would produce 40,000 gallons of cellulosic ethanol a year at a commercial demonstration plant near Pittsburgh (Green Car Congress Website, 2008). Coskata's process varies significantly from the traditional biochemical process, with ethanol produced via the fermentation of synthetic gas, or 'syngas', mainly made up of carbon monoxide and hydrogen. This process offers the largest flexibility as a wide assortment of raw materials can be broken down into syngas using the gasification process. A life cycle analysis performed by Argonne National Laboratory has determined that the Coskata process yields up to a 7.7 net energy balance, which compares favourably to the 1.3 net energy balances reported for corn-based ethanol. Water usage for the process is forecasted to be as low as 1 gallon or less per gallon of ethanol produced versus 3-6 for the corn process. It is claimed by Coskata Inc. that the Coskata process can produce ethanol at under US $1 / gallon.

In comparison to ethanol, butanol (C_4H_9OH) has a number of advantages for use as a biofuel. It is more hydrophobic; has a higher energy density; can be transported through existing pipeline infrastructure; and can be mixed with gasoline at any ratio. With properties superior to that of ethanol, butanol is generally being considered as a gasoline blend component that could be used in higher concentrations. DuPont-BP venture is using existing technology to convert sugar beets into 30,000 tons, or 9 million gallons, of biobutanol annually at British Sugar's facility in Wissington, England, east of Cambridge.(BP Website, 2006) The BP-DuPont partners have demonstrated 16% blend rates with biobutanol versus a 10% blend rate with ethanol (Bu16 vs. E10). Test results presented by BP and DuPont showed that bio-derived 1-butanol (also called n-butanol) performs similarly to unleaded gasoline on key parameters, and that biobutanol formulations meet key characteristics of a "good" spark ignition fuel, including high energy density, controlled volatility, sufficient octane and low levels of impurities.

Up to this point, not much study has been carried out in terms of using microalgae or microalgael biomass for ethanol production. However, it was reported that a few microalgael species could accumulate large quantities of starch as storage materials. The microalgael starch could be employed for ethanol production via saccharification and fermentation similar to ethanol production from corn. It is also possible to use the hybrid approach for ethanol production of ethanol, i.e., gasify microalgael biomass to produce syngas and then produce ethanol by fermentation.

4.4. Bio-Oil and Bio-Syngas

When biomass is processed under high temperature in the absence of oxygen, products are produced as three phases: the vapour phase, the liquid phase, and the solid phase. The liquid phase is a complex mixture called bio-oil. Extensive studies have been carried out on biomass conversion; several technologies such as entrained flow reactor, circulating fluid bed gasifier (Meier and Faix 1999; Prins, Ptasinski et al. 2006; Zwart, Boerrigter et al. 2006), vacuum pyrolysis (Pakdel and Roy 1991), and vortex reactor (Meier and Faix 1999) have been demonstrated to be effective. The overall energy to biomass ratio of a well controlled pyrolytic process could be as high as 95.5% (Demirbas 2001). These technologies can be classified into two categories: 1) Pyrolysis, the primary product of which is pyrolytic liquids (bio-oils) and 2) Gasification, with "syngas" as the primary product (Prins, Ptasinski et al. 2006; Lv, Yuan et al. 2007). Canadian companies have shown world leadership in this field. Ensyn Corp (EC), a private company based in Ottawa, Ontario, developed one of the most successful biomass conversion technologies, Rapid Thermal Processing. RTP™ is a patented, state-of-the-art process that transforms carbon-based feedstocks, either wood "biomass" or petroleum hydrocarbons, into more valuable chemical and fuel products. Plascoenergy Group, another private company based in Ottawa, Ontario, has a proprietary Plasco Conversion System (PCS) that converts carbonaceous materials such as municipal solid waste, into an energy-rich fuel or "syngas" and a commercially useful, inert solid, or "slag".

Both bio-oils have been demonstrated to be suitable for power generation via both external combustion (e.g., steam cycles, organic Rankine cycles, and Stirling engines) and internal combustion (e.g., diesel engines and gas-turbine engines) or by co-firing with fossil diesel or natural gas (Bridgwater, Meier et al. 1999; Czernik and Bridgwater 2004; Chiaramonti, Oasmaa et al. 2007). Nevertheless, they have several undesirable features such as high oxygen content, low heat content, high viscosity at low temperature, and chemical instability (Bridgwater, Meier et al. 1999; Czernik and Bridgwater 2004; Chiaramonti, Oasmaa et al. 2007) that impede their use as quality transportation fuels. Production of high-grade transportation fuels from biomass has been demonstrated to be technically feasible by gasification and subsequent Fischer Tropsch synthesis (Petrus and Noordermeer 2006; Lv, Yuan et al. 2007). Recent work by a group in China has demonstrated that hydrogen gas can be derived reliably by steam-reforming bio-oil (Wang, Pan et al. 2007).

Most studies have so far focused on the use of conventional biomasses from forestry and agricultural sources (Demirbas 2001). It was estimated that in year 2000, the majority of biomass energy was produced from wood and wood wastes (64%), followed by municipal solid waste (MSW) (24%), agricultural waste (5%) and landfill gases (5%) (Demirbas and industry 2000; Demirbas 2001). Recently, a few investigations have been carried out regarding the suitability of microalgael biomass for bio-oil production (Miao and Wu 2004; Miao, Wu et al. 2004; Demirbas 2006). It was shown that, in general, microalgae bio-oils have higher quality than bio-oil from wood (Demirbas 2006).

4.5. THERMOCHEMICAL CONVERSION OF BIOMASS

There are several ways to convert microalgae biomass to biofuels, which can be classified into biochemical conversion, chemical reaction, direct combustion and thermochemical conversion (Fig. 4.3) (Demirbas 2001; McKendry 2002). More specifically, example processes belonging to biochemical conversion include anaerobic digestion for methane production and fermentation for ethanol production (Spolaore, Joannis-Cassan et al. 2006); an example chemical conversion process involves extraction of lipids accumulated in microalgae cells and conversion of the extracted lipid to biodiesel via a simple transesterification reaction (Belarbi, Molina et al. 2000; Chisti 2007); and some example thermochemical conversion processes include pyrolysis (Chiaramonti, Oasmaa et al. 2007), gasification (Hirano, Hon-Nami et al. 1998) and liquefaction (Minowa and Sawayama 1999). Energy stored in microalgael biomass could also be utilized via direct combustion or co-firing. Thermochemical conversion (Demirbas 2004) is one of the most practical biomass conversion strategies.

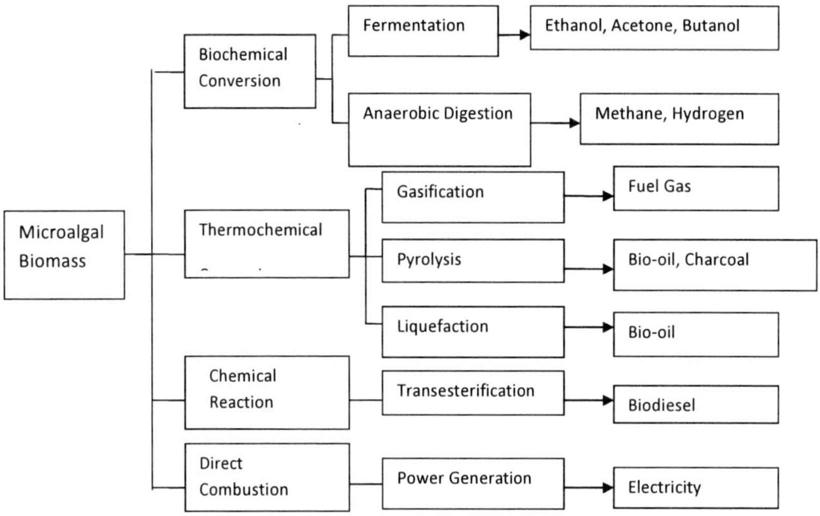

Figure 4.3. Energy production via microalgael biomass conversion using biochemical, thermochemical, chemical and direct combustion processes (Wang B., et al., 2008).

4.5.1. Gasification

Gasification is the conversion of biomass into combustible gas mixture by the partial oxidation of biomass at high temperatures, typically in the range of 800 - 900 °C (Elliott and Sealock Jr 1996; McKendry 2002). The low calorific value gas produced (about 4-6 MJ N m^{-3}) can be burnt directly for heating or electricity generation or used as a fuel for engines and gas turbines. According to the operation conditions and mechanisms involved, gasification processes can be classified into two categories: conventional gasification and supercritical water gasification.

Conventional gasification is to decompose dry biomass at high temperature, pressure and the absence of oxygen to tar materials, which are further decomposed into small molecular combustible gas, usually with the help of gasification catalysts. Supercritical water gasification, on the other hand, relies on the existence of supercritical water to cause the hydrolysis of biomass components to produce smaller molecules (Feng, van der Kooi et al. 2004; Matsumura, Sasaki et al. 2006). Due to the different reaction mechanisms, supercritical water gasification owns some unique advantages over conventional gasification. Firstly, supercritical water gasification is suitable for recovery of energy from wet biomass, avoiding the energy intensive drying process. Secondly, supercritical water has some specific features such as high solubility of biomass components and products, which could achieve homogeneous reaction and allows simple separation of the gas products from liquid phase at the end of reaction (Feng, van der Kooi et al. 2004). Thirdly, supercritical water gasification requires more moderate conditions than that of conventional gasification. Typically, to get fuel gas by supercritical water gasification, reaction temperature around 620 K (347 °C) is used when utilizing metal catalyst, while temperature above 970 K (697 °C) with alkali or carbonaceous catalysts can achieve complete gasification (Sutton, Kelleher et al. 2001). Both of these two reaction temperatures are much lower than that required by conventional gasification (see previous discussion). It is evident that the low operation temperature is desirable from the standpoint of reactor cost. Due to the clear advantages associated with supercritical water gasification, extensive studies has been taken to develop supercritical water gasification process of biomass with a high moisture content (Elliott and Sealock Jr 1996; Antal Jr, Allen et al. 2000; Sutton, Kelleher et al. 2001; Osada, Sato et al. 2004).

4.5.2. Pyrolysis

Pyrolysis is the conversion of dry biomass to liquid (termed bio-oil or bio-crude), solid and gaseous fractions, by heating the biomass in absence of oxygen to around 500℃. If the purpose is to maximize the yield of liquid products resulting from biomass pyrolysis, a low temperature, high heating rate, short gas residence time process would be required (Demirbas 2000; Demirbas 2001). Pyrolysis can be used to produce predominantly bio-oil if flash pyrolysis is used, enabling the conversion of biomass to bio-crude with an efficiency of up to 80% (McKendry 2002). Similar to conventional gasification, pyrolysis also requires the feedstock to be dry biomass, requiring the energy intensive drying process.

The bio-oil produced from pyrolysis can be combusted directly in engines and turbines to produce energy. However, problems with its poor thermal stability still need to be overcome. Upgrading bio-oils can be achieved by lowering the oxygen content and removing alkalis by means of hydrogenation and catalytic cracking of the oil (Chiaramonti, Oasmaa et al. 2007).

A few studies have been carried out on the production of fuel oil from microalgae by pyrolysis (Bayer 1981; Miao and Wu 2004; Miao, Wu et al. 2004). Yields of 18% and 24% high quality bio-oil were obtained by fast pyrolysis of *Chlorella protothecoides* and *Microcystis aeruginosa* at temperature of 500°C with a heating rate of 600 °C s^{-1} (Miao, Wu et al. 2004), which has a potential for commercial application of large-scale production of liquid fuels. It was also noticed (Peng, Wu et al. 2001) that *Chlorella protothecoides* (CP) was preferable for pyrolysis over *Spirulina platensis* (SP) (Spolaore, Joannis-Cassan et al.) because the value of activation energy for CP pyrolysis was $4.22\text{-}5.25\times10^4$, lower than that of SP, which was $7.62\text{-}9.70\times10^4$. Furthermore, the char in final residue of CP was 14.00-15.14%, less than that of SP by 2-3%.

4.5.3. Liquefaction

Liquefaction is a process to obtain liquid fuels by thermo-chemical conversion of biomass at low temperature and high pressure using a catalyst in the presence of hydrogen. It has the advantage of treating wet materials (water content more than 60%) compared with direct combustion, conventional gasification and pyrolysis. Since it requires no drying process, microalgae, which typically have high moisture contents, are a good raw material for the liquefaction. However, it is worth noting that liquefaction is a relatively expensive process due to the use of hydrogen. In addition, the product is a tarry lump, which is difficult to handle (Goyal, Seal et al. 2008).

Studies have been carried out on the direct thermochemical liquefaction of biomass of some microalgae species. *Botryococcus braunii* (Minowa and Sawayama 1999), which is a colony-forming green microalgae and capable of accumulating hydrocarbons to 30-70% of its dry weight, was effectively converted to liquid fuel by liquefaction. Earlier on, *B. braunii* had been suggested to be considered as a prime candidate for biological carbon dioxide fixation combined with liquefaction to produce liquid fuel (Dote, Sawayama et al. 1994). Using CO_2 as the carbon source, a growth rate of 1.0 g l^{-1} was obtained with *Dunaliella* under non-sterilized conditions after one week of cultivation (0.17 g $l^{-1}day^{-1}$) (Kishimoto, Okakura et al. 1994). The yield of its conversion to oil by liquefaction was 36%, indicating the capacity for efficient oil production and CO_2 fixation. Furthermore, *Dunaliella* is a green halophilic algae that could grow under highly salinity conditions (6% of NaCl aqueous solution) without the need of sterilization. This would be a great advantage in large-scale cultivation for CO_2 mitigation. In addition, *Dunaliella* can accumulate ß-carotene to more than 10% of its dry cell weight, providing a commercially profitable route for microalgael CO_2 bio-mitigation. Since the technology for the mass cultivation of *Dunaliella* has been established, it would be advantageous to use *Dunaliella* for combined biofuel production, CO_2 mitigation, and novel bioproduct (β-carotene) production.

REFERENCES

Antal Jr, M. J., S. G. Allen, et al. (2000). "Biomass gasification in supercritical water." *Industrial and Engineering Chemistry Research* 39(11): 4040-4053.

Bayer, F. L. (1981). *Pyrolysis gas chromatographic characterization differentiation and identification of biopolymers - an overview*. American Chemical Society, Polymer Preprints, Division of Polymer Chemistry.

Belarbi, E. H., E. Molina, et al. (2000). "A process for high yield and scaleable recovery of high purity eicosapentaenoic acid esters from microalgae and fish oil." *Enzyme and Microbial Technology* 26(7): 516-529.

BP Website. 2006. BP-DuPont biofuels fact sheet.

Bridgwater, A. V., D. Meier, et al. (1999). "An overview of fast pyrolysis of biomass." *Organic Geochemistry* 30(12): 1479-1493.

Bridgwater, A. V., A. J. Toft, et al. (2002). "A techno-economic comparison of power production by biomass fast pyrolysis with gasification and combustion." *Renewable and Sustainable Energy Reviews* 6(3): 181-248.

Canada, N. R. (2006). *Energy Use Data Handbook*, 1990 and 1998 to 2004 N. R. Canada.

Chiaramonti, D., A. Oasmaa, et al. (2007). "Power generation using fast pyrolysis liquids from biomass." *Renewable and Sustainable Energy Reviews* 11(6): 1056-1086.

Chisti, Y. (2007). "Biodiesel from microalgae." *Biotechnology Advances* 25(3): 294-306.

Crutzen, P. J., A. R. Mosier, et al. (2007). "N2O release from agro-biofuel production negates global warming reduction by replacing fossil fuels." *Atmospheric Chemistry and Physics Discussions* 7(4): 11191-11205.

Czernik, S. and A. V. Bridgwater (2004). "Overview of applications of biomass fast pyrolysis oil." *Energy and Fuels* 18(2): 590-598.

Das, D. and T. N. Veziro?lu (2001). "Hydrogen production by biological processes: A survey of literature." *International Journal of Hydrogen Energy* 26(1): 13-28.

Das, D. and T. N. Veziroglu (2008). "Advances in biological hydrogen production processes." *International Journal of Hydrogen Energy* 33(21): 6046-6057.

De la Noue, J., De Pauw, N. (1988). "The Potential of Microalgael Biotechnology: a review of production and uses of microalgae." *Biotechnology Advances* 6: 725-770.

Demirbas, A. (2000). "Mechanisms of liquefaction and pyrolysis reactions of biomass." *Energy Conversion and Management* 41(6): 633-646.

Demirbas, A. (2001). "Biomass resource facilities and biomass conversion processing for fuels and chemicals." *Energy Conversion and Management* 42(11): 1357-1378.

Demirbas, A. (2004). "Current technologies for the thermo-conversion of biomass into fuels and chemicals." *Energy Sources* 26(8): 715-730.

Demirbas, A. (2006). "Oily products from mosses and algae via pyrolysis." *Energy Sources, Part A: Recovery, Utilization and Environmental Effects* 28(10): 933-940.

Demirbas, A. (2007). "Recent developments in biodiesel fuels." *International Journal of Green Energy* 4(1): 15-26.

Demirbas, A. and B. r. f. e. a. c. industry (2000). "Biomass resources for energy and chemical industry." *Energy Edu Sci Technol* 5: 21-45.

Dote, Y., S. Sawayama, et al. (1994). "Recovery of liquid fuel from hydrocarbon-rich microalgae by thermochemical liquefaction." *Fuel* 73(12): 1855-1857.

Dube, M. A., A. Y. Tremblay, et al. (2007). "Biodiesel production using a membrane reactor." *Bioresource Technology* 98(3): 639-647.

Elliott, D. C. and L. J. Sealock Jr (1996). "Chemical processing in high-pressure aqueous environments: Low-temperature catalytic gasification." *Chemical Engineering Research and Design* 74(5): 563-566.

Feng, W., H. J. van der Kooi, et al. (2004). "Phase equilibria for biomass conversion processes in subcritical and supercritical water." *Chemical Engineering Journal* 98(1-2): 105-113.

Gaffron, H. and J. Rubin (1942). "Fermentative and photochemical production of hydrogen in algae." *J Gen Physiol* 28: 269-285.

Goyal, H. B., D. Seal, et al. (2008). "Bio-fuels from thermochemical conversion of renewable resources: A review." *Renewable and Sustainable Energy Reviews* 12(2): 504-517.

Green Car Congress Website. 2008. *Coskata Chooses Site for Demo Syngas-to-Ethanol Plant*, Green Car Congress. 25 April 2008

Hirano, A., K. Hon-Nami, et al. (1998). "Temperature effect on continuous gasification of microalgael biomass: Theoretical yield of methanol production and its energy balance." *Catalysis Today* 45(1-4): 399-404.

Huntley, M. E. and D. G. Redalje (2007). "CO2 mitigation and renewable oil from photosynthetic microbes: A new appraisal." *Mitigation and Adaptation Strategies for Global Change* 12(4): 573-608.

J Sheehan, e. a. (1998). *A look back at the US Department of Energy's aquous species program: biodiesel from Algae,* The National Renewable Energy Laboratory, US Department of Energy.

Kishimoto, M., T. Okakura, et al. (1994). "CO2 fixation and oil production using micro-algae." *Journal of Fermentation and Bioengineering* 78(6): 479-482.

Kondili, E. M. and J. K. Kaldellis (2007). "Biofuel implementation in East Europe: Current status and future prospects." *Renewable and Sustainable Energy Reviews* 11(9): 2137-2151.

Lewis, N. S. and D. G. Nocera (2006). "Powering the planet: Chemical challenges in solar energy utilization." *Proceedings of the National Academy of Sciences of the United States of America* 103(43): 15729-15735.

Li, X., H. Xu, et al. (2007). "Large-scale biodiesel production from microalgae Chlorella protothecoides through heterotrophic cultivation in bioreactors." *Biotechnology and Bioengineering* 98(4): 764-771.

Lv, P., Z. Yuan, et al. (2007). "Bio-syngas production from biomass catalytic gasification." *Energy Conversion and Management* 48(4): 1132-1139.

Marchetti, J. M., V. U. Miguel, et al. (2007). "Possible methods for biodiesel production." *Renewable and Sustainable Energy Reviews* 11(6): 1300-1311.

Matsumura, Y., M. Sasaki, et al. (2006). "Supercritical water treatment of biomass for energy and material recovery." *Combustion Science and Technology* 178(1-3): 509-536.

McCarthy S. *Iogen on track to get funding for biofuel refiner.* The Globe and Mail, Sat 15 Mar 2008.

McKendry, P. (2002). "Energy production from biomass (part 3): Gasification technologies." *Bioresource Technology* 83(1): 55-63.

Meher Kotay, S. and D. Das (2008). "Biohydrogen as a renewable energy resource-Prospects and potentials." *International Journal of Hydrogen Energy* 33(1): 258-263.

Meher, L. C., D. Vidya Sagar, et al. (2006). "Technical aspects of biodiesel production by transesterification - A review." *Renewable and Sustainable Energy Reviews* 10(3): 248-268.

Meier, D. and O. Faix (1999). "State of the art of applied fast pyrolysis of lignocellulosic materials - A review." *Bioresource Technology* 68(1): 71-77.

Miao, X. and Q. Wu (2004). "High yield bio-oil production from fast pyrolysis by metabolic controlling of Chlorella protothecoides." *Journal of Biotechnology* 110(1): 85-93.

Miao, X. and Q. Wu (2006). "Biodiesel production from heterotrophic microalgael oil." *Bioresource Technology* 97(6): 841-846.

Miao, X., Q. Wu, et al. (2004). "Fast pyrolysis of microalgae to produce renewable fuels." *Journal of Analytical and Applied Pyrolysis* 71(2): 855-863.

Minowa, T. and S. Sawayama (1999). "Novel microalgael system for energy production with nitrogen cycling." *Fuel* 78(10): 1213-1215.

Nakamura, D. N. (2006). "Journally speaking: The mass appeal of biomass." *Oil and Gas Journal* 104(45): 15.

Osada, M., T. Sato, et al. (2004). "Low-temperature catalytic gasification of lignin and cellulose with a ruthenium catalyst in supercritical water." *Energy and Fuels* 18(2): 327-333.

Pakdel, H. and C. Roy (1991). "Hydrocarbon content of liquid products and tar from pyrolysis and gasification of wood." *Energy & Fuels* 5(3): 427-436.

Peng, W., Q. Wu, et al. (2001). "Pyrolytic characteristics of microalgae as renewable energy source determined by thermogravimetric analysis." *Bioresource Technology* 80(1): 1-7.

Petrus, L. and M. A. Noordermeer (2006). "Biomass to biofuels, a chemical perspective." *Green Chemistry* 8(10): 861-867.

Prins, M. J., K. J. Ptasinski, et al. (2006). "More efficient biomass gasification via torrefaction." *Energy* 31(15): 3458-3470.

Renewable Fuels association. (2008). *U.S. Cellulosic Ethanol Projects Under Development and Construction*, September 2008

Richmond, A. (2000). "Microalgael biotechnology at the turn of the millennium: A personal view." *Journal of Applied Phycology* 12(3): 441-451.

Seefeldt, L. C. (2007). "Utah group plans to make biodiesel from algae." *Industrial Bioprocessing* 29(3): 5-6.

Sigfusson, T. I. (2007). "Pathways to hydrogen as an energy carrier." *Philosophical Transactions of the Royal Society A: Mathematical, Physical and Engineering Sciences* 365(1853): 1025-1042.

Spolaore, P., C. Joannis-Cassan, et al. (2006). "Commercial applications of microalgae." *Journal of Bioscience and Bioengineering* 101(2): 87-96.

Sutton, D., B. Kelleher, et al. (2001). "Review of literature on catalysts for biomass gasification." *Fuel Processing Technology* 73(3): 155-173.

Tornabene, T. G., G. Holzer, et al. (1983). "Lipid Composition of the Nitrogen Starved Green Algae Neochloris Oleoabundans." *Enzyme and Microbial Technology* 5(6): 435-440.

UN (2003). *World Energy Assessment Report: Energy and the Challenge of Sustainability*. . NY, United Nations Development Program United Nations.

W. Barkley, C. W., R.A. Lewin, L. Cheng (1987). *Development of microalgael systems for the production of liquid fuels.* Société pour l'Algologie Appliquée, Villeneuve d'Ascq, France, Elsevier Applied Science.

Wang, Z., Y. Pan, et al. (2007). "Production of hydrogen from catalytic steam reforming of bio-oil using C12A7-O--based catalysts." *Applied Catalysis A: General* 320: 24-34.

Xu, H., X. Miao, et al. (2006). "High quality biodiesel production from a microalgae Chlorella protothecoides by heterotrophic growth in fermenters." *Journal of Biotechnology* 126(4): 499-507.

Zwart, R. W. R., H. Boerrigter, et al. (2006). "The impact of biomass pretreatment on the feasibility of overseas biomass conversion to Fischer-Tropsch products." *Energy and Fuels* 20(5): 2192-2197.

Chapter 5

ENHANCEMENT OF LIPID PRODUCTION USING BIOCHEMICAL, GENETIC AND TRANSCRIPTION FACTOR ENGINEERING APPROACHES*

Dorval Courchesne N.M. Dorval, Albert Parisien, Bei Wang and Christopher Q. Lan
The University of Ottawa

5.1. INTRODUCTION

As discussed in Chapter 4, biodiesel is one of the most promising renewable transportation fuels that have achieved remarkable success worldwide. According to a World Bank report (2008), 6.5 billion litres of biodiesel was produced worldwide in 2006, 75% of which by the European Union and 13% by the USA. The current contribution of biodiesel to global transportation fuel consumption is, however, only 0.14% and the favourable policies of major countries in the world are expected to increase this contribution by five times by 2020. It is therefore predictable that massive global demand on renewable energy will continue to drive the rapid growth of biodiesel production in an unprecedented scale. Nevertheless, current increase of food prices worldwide had brought about public

* This chapter is based on an article with the same title, which was published in Journal of Biotechnology 141: 31-41.

awareness and concerns regarding the competition for agricultural resources between the food industry and the energy sector. Development of sustainable and cost-effective alternatives to the traditional agricultural and forestry crops is therefore of urgent need for sustainable biofuel production.

Oil-rich microalgae have been demonstrated to be a promising alternative source of lipids for biodiesel production (Chisti 2007; Li et al. 2008b; Song et al. 2008; Walker et al. 2005b; Wang et al. 2008). There seems to be little doubt that fast growing microalgae should be able to provide enough renewable biofuels for the replacement of fossil transportation fuels (Li et al. 2008b). An integrated strategy was proposed to enhance the economical cost-effectiveness and environmental sustainability by combining the benefits of biofuel production, CO2 mitigation, waste heat utilization, wastewater treatment and novel bioproduct production using the microalgael cultivation processes (Li et al. 2008b; Wang et al. 2008). Nevertheless, significant challenges remain in the economics of microalgael biodiesel production and extensive studies have been carried out to cope with these challenges. In this chapter, we discuss the global lipid biosynthesis pathway and compare three possible strategies for lipid overproduction in microalgae: the Biochemical Engineering (BE) approaches, the Genetic Engineering (GE) approaches, and the Transcription Factor Engineering (TFE) approaches.

5.2. An Overview of the Global Lipid Biosynthesis Pathway

As shown in Figure 5.1, the global synthesis pathway of TAG in cells is comprised of three major steps: 1) carboxylation of acetyl-CoA to form malonyl-CoA, the committing step of fatty acid biosynthesis; 2) acyl chain elongation; and 3) TAG formation. The enzymes involved in each step of the pathway and their functionalities are discussed briefly as follows.

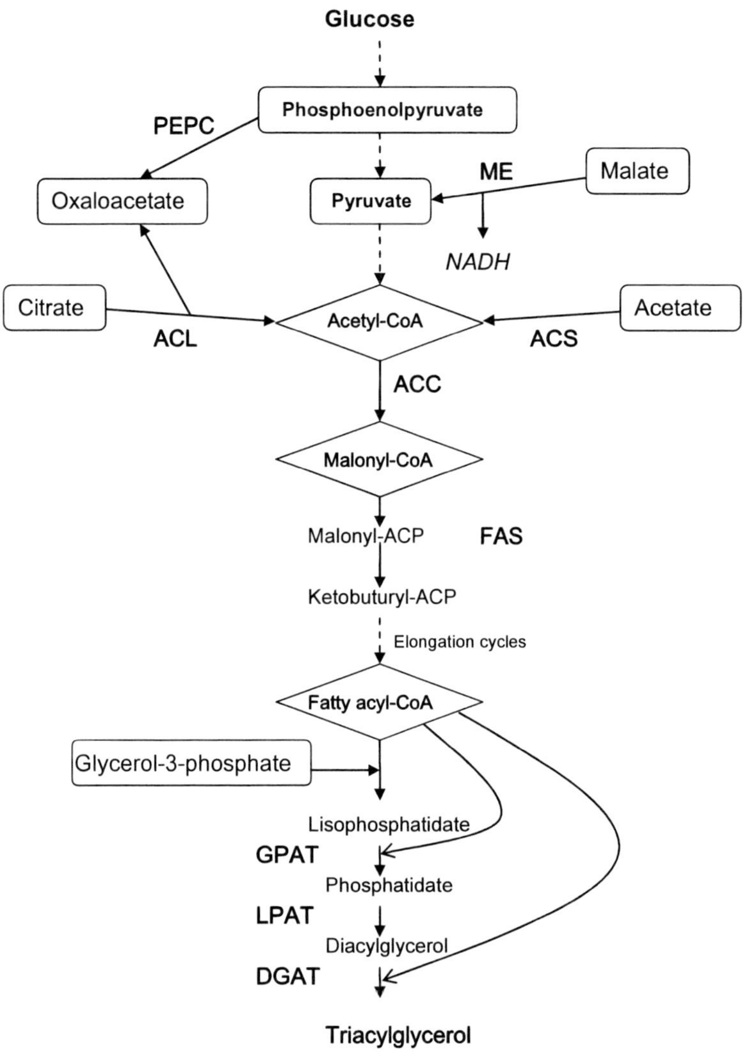

Figure 5.1. The fatty acid and TAG biosynthesis.

5.2.1. The Committing Step

As shown in Figure 5.1, lipid biosynthesis starts with the Acetyl-CoA Carboxylase (ACC), which catalyzes the important committing step of the fatty acid synthetic pathway, the biotin-dependant carboxylation of acetyl-CoA to form malonyl-CoA (Davis et al. 2000; Kim 1997; Li and Cronan Jr 1993; Sendl et al. 1992). In *Escherichia coli*, ACC is a protein containing four subunits, which are encoded by genes accA, accB, accC and accD that are located at different positions on the chromosome (Li and Cronan Jr 1993). It is a trifunctional enzyme with a biotin carboxyl carrier protein, a biotin-carboxylase subunit and a carboxyl-transferase subunit (Sendl et al. 1992) joined together into a heterotrimeric complex (Tehlivets et al. 2007). In contrast, eukaryotic cells encode a multi-domain single polypeptide, which is responsible for all the functions of the ACC (Sasaki and Nagano 2004; Tehlivets et al. 2007). In animal cells, ACC is located in the cytoplasm and thus has to use cytosolic acetyl-CoA for malonyl-CoA formation and acyl chain elongation. Yeasts have both cytosolic and mitochondrial ACC, but it has been demonstrated to be able to survive with a non-functional mitochondrial enzyme (Tehlivets et al. 2007). In plants, fatty acid synthesis occurs entirely in plastids of developing seeds, and ACC uses the acetyl-CoA that is found in this organelle (Dyer and Mullen 2005; Roesler et al. 1997). The plastid ACC has a different structure than the cytosolic ACC. It is a multi-subunit prokaryotic type enzyme, as opposed to the multifunctional eukaryotic type located in the cytoplasm.

5.2.2. Acyl Chain Elongation

Once malonyl-CoA is synthesized, it is transferred by malonyl-CoA:ACP transacetylase to the acyl-carrier protein (ACP) of the fatty acid synthase (FAS) multi-enzymatic complex (Subrahmanyam and Cronan Jr 1998). Bacteria and plants have type II FAS (Rock and Jackowski 2002), which is a multi-subunit protein in which each individual peptide is dissociable and can catalyze an enzymatic reaction, as opposed to the type I FAS found in yeast and vertebrates, which is a multifunctional protein (Verwoert et al. 1995).

FAS catalyzes fatty acid elongation by condensing malonyl-CoA molecules and acetyl-CoA. ACP, one of the FAS subunits, contains a thiol group that can form malonyl-ACP via forming thioesters with malonyl-CoA, and afterwards with the growing acyl chain in order to assure its transport (Subrahmanyam and Cronan Jr 1998). ACP can also fix acetyl by forming acetyl-ACP. Then, the acetyl-group is transferred to another subunit of the FAS, the ketoacyl-ACP synthase (KAS), which catalyzes the condensation of malonyl-ACP or the growing acyl chain to form ketobutyryl-ACP or ketoacetyl-ACP. This resulting compound is first transformed via three successive reactions, i.e., reduction, dehydration and reduction, and then condensed with another malonyl-CoA. This cycle is repeated until the saturated chain of a palmitic (16:0) or a stearic acid (18:0) is formed (Subrahmanyam and Cronan Jr 1998). At last, ACP-thioesterase cleaves the acyl chain and liberates the fatty acid.

To obtain longer or unsaturated chains, elongases and desaturases are required, which act on palmitate or stearate. These enzymes are located in endoplasmic reticulum membrane and mitochondria. They can produce long chain fatty acids, as well as unsaturated acyl chains. They then act on the composition of the fatty acid pool but not on their accumulation level. Many experiments have been carried out to modify the lipid content in transgenic plants using these enzymes, such as the increase of omega-3 production (Budziszewski et al. 1996; Dehesh 2001; Graham et al. 2007; Ivy et al. 1998; Napier 2007; Napier et al. 2004; Opsahl-Ferstad et al. 2003; Stoll et al. 2005; Zou et al. 1997).

5.2.3. Triacylglycerol (TAG) Formation

For eukaryotes, TAG formation takes place in specialized organelles, i.e., the mitochondria or/and plastid (plants only) located in the endoplasmic reticulum. In contrast, the TAG synthesis takes place in the cytoplasm of prokaryotic cells. This process yields neutral lipids, a way to store fatty acids and thus energy (Rajakumari et al. 2008). Storage of high energy density TAG allows cells to have more free space (Coleman and Lee 2004).

The first step of TAG synthesis is the condensation (acylation) of Glycerol-3-phosphate (G3P) with an acyl-CoA to form lysophosphatidate (LPA), which is catalyzed by acyl-CoA:glycerol-*sn*-3-phosphate acyl-transferase (GPAT). This enzyme exhibits the lowest specific activity of the TAG synthesis pathway, and was suggested to be potentially the rate limiting step (Cao et al. 2006; Coleman and Lee 2004). It is subjected to many regulatory controls at the transcriptional level, at the post-transcriptional level (e.g., by means of post-transcriptional phosphorylation or dephosphorylation) and by allostery.

The LPA is then further condensed, catalyzed by acyl-CoA:acylglycerol-*sn*-3-phosphate acyl-transferase (GPAT), with another acyl-CoA to produce phosphatidate (PA) (Athenstaedt and Daum 1999). Afterwards, PA can be dephosphorylated by phosphatidic acid phosphatase (PAP) to produce diacylglycerol.

At last, synthesis of TAG is catalyzed by acyl-CoA:diacylglycerol acyl-transferase (DGAT), which incorporates the third acyl-CoA into the diacylglycerol molecule. This enzyme is also known as an important regulator for this pathway (Oelkers et al. 2002; Sandager et al. 2002). TAGs can then be stored in oil bodies (Murphy 2001).

5.3. Lipid Production Enhancement: The Biochemical Engineering (BE) Approaches

The BE approach here refers to the strategy of enhancing lipid production of microalgae by controlling the nutritional or cultivation conditions (e.g. temperature, pH, and salinity) to channel metabolic flux generated in photobiosynthesis into lipid biosynthesis. Nutrient starvation has so far been the most commonly employed approach for directing metabolic fluxes to lipid biosynthesis of microalgae. In this scenario, microalgae accumulate lipids as a means of storage under nutrient limitation when energy source (i.e. light) and carbon source (i.e. CO_2) are abundantly available and when the cellular mechanisms for the photobiosynthesis are active. While a number of nutrients such as phosphorus and iron deficiency have been reported as being able to cause cell growth cessation and channel metabolic flux to lipid/fatty acid biosynthesis, nitrogen is the most commonly reported nutritional limiting factor triggering lipid accumulation in microalgae.

Nitrogen starvation has been observed to lead to lipid accumulation in a number of microalgael species. For instance, *Chlorella* usually accumulates starch as storage material. However, it was observed by Illman et al. (2000) that *C. emersonii*, *C. minutissima*, C. *vulgaris*, and *C. pyrenoidosa* could accumulate lipids of up to 63%, 57%, 40%, and 23% of their cells on a dry weight basis, respectively, in low-N medium. *Neochloris oleoabundans* was reported, under nitrogen deficient conditions, to be able to accumulate 35-54% lipids of its cell dry weight and its TAGs comprised 80% of the total lipids (Kawata et al. 1998; Tornabene et al. 1983). It was also observed (Yamaberi et al. 1998) that the TAGs accumulated in *Nannochloris sp.* cells could be 2.2 times as that in the cells in nitrogen sufficient cultures. Our studies (Li et al. 2008a) showed that sodium nitrate was the most favourable nitrogen source for cell growth and lipid production of *N. oleoabundans* among the three tested nitrogen-containing compounds, i.e., sodium nitrate, urea, and ammonium bicarbonate. It was observed that lipid cell contents decreased with the increase of sodium nitrate in the medium in the range of 3 mM to 20 mM. The trend that lower nitrogen source concentration in medium led to higher lipid cell content was hypothetically explained by the fact that nitrogen would have exhausted earlier at low cell density when the initial concentration of nitrogen source in medium was low. As a result, cells started to accumulate lipid when light had good penetration (at low cell density), when individual cells were exposed to a large quantity of light energy, resulting in more metabolic flux generated from photosynthesis to be channelled to lipid accumulation on an unit biomass basis.

Phosphate limitation was also observed to cause enhancement of lipid accumulation of *Monodus subterraneus* (Khozin-Goldberg and Cohen 2006). With decreasing phosphate availability from 175 to 52.5, 17.5 and 0 μM (K_2HPO_4), the cellular total lipid content of starved cells increased, mainly due to the drastic increase in TAG levels. In the absence of phosphate, the proportion of phospholipids was reduced from 8.3% to 1.4% of total lipids, and the proportion of TAG increased from 6.5% up to 39.3% of total lipids. Furthermore, iron deficiency has also been reported to stimulate lipid accumulation in microalgae *Chlorella vulgaris*, which accumulated up to 56.6% lipid of biomass by dry weight under the optimal condition (1.2×10^{-5} mol $FeCl_3$) (Liu et al. 2008).

In addition to nutrient-starvation, other stress conditions may also cause enhanced accumulation of lipids in microalgae. For instance, Takagi et al. (2006) observed that TAG content increased in *Dunaliella*, a marine algae, under high salinity conditions. In that research, an initial NaCl concentration higher than 1.5 M was found to markedly inhibit cell growth. However, when the initial NaCl concentration increased from 0.5 (equal to seawater) to 1.0 M, it resulted in a higher intracellular lipid content (67%) in comparison with 60% for the salt concentration of 0.5 M. Addition of 0.5 or 1.0 M NaCl at mid-log phase or the end of log phase during cultivation with initial NaCl concentration of 1.0 M further increased the lipid content to 70%.

An inherited disadvantage of the BE strategy is, however, nutrient starvation or the physiological stress required for accumulating high lipid content in cells is associated with reduced cell division (Ratledge 2002). Since lipids are intracellular products, the overall lipid productivity is the product of cell lipid content multiplied by biomass productivity. The overall lipid/energy productivity will therefore be compromised due to the lowered biomass productivity. For instance, Scagg et al. (2002) studied the energy recovery of *Chlorella vulgaris* and *C. emersonii* grown in Watanabe's medium and a low nitrogen medium. The results showed that the low nitrogen medium, although induced higher lipid accumulation in both algae with high calorific values, the overall energy recovery was lower with the low nitrogen medium than that with the Watanabe's medium. Our studies (Li et al. 2008a) also showed that, in the tested range of 3 mM to 20 mM sodium nitrate, although the highest cell lipid content of 40% was obtained at the lowest sodium nitrate concentration of 3 mM, the maximal lipid productivity was achieved at 5 mM.

A commonly suggested countermeasure is to use a two-stage cultivation strategy, dedicating the first stage for cell growth/division in nutrient-sufficient medium and the second stage for lipid accumulation under nutrient-starvation or other physiological stress. Indeed, a well formulated medium as the one proposed by our group in a previous study (Li et al. 2008a) would achieve the two-stage lipid production "naturally" as the cells will be able to grow quickly before the exhaustion of the limiting substrate (N in this particular case) and then switch to lipid accumulation under N-starvation conditions. Furthermore, a hybrid closed photobioreactor/open pond microalgael cultivation system (Huntley and Redalje 2007) was suggested to be potentially the appropriate engineering solution accommodating the two-stage strategy with the photobioreactors dedicated to nutrient-rich inoculum build-up and the open ponds to low-nutrient lipid accumulation (Schenk et al. 2008). It was also pointed out that employment of low-nutrient media in open ponds is not only beneficial for lipid accumulation and contamination control, but also environmentally friendly.

Nevertheless, deficiency of these nutrients may slow down photosynthesis of microalgael cells one way or the other, resulting in lowered overall lipid productivity. Many of the commonly used limiting nutrients are essential for photosynthesis of microalgae and the depletion of which may severely impede the photosynthesis responsible for generating the metabolic flux for lipid production. For instance, it was observed in our studies that chlorophyll, the essential pigment for light capturing in the biosynthesis of green algae *N. oleoabundans*, was consumed for cell growth when nitrogen was exhausted from the medium, resulting in a sharp drop of chlorophyll cell content (Li et al. 2008a). Phosphorus is essential to the cellular processes related to energy bio-conversion (e.g., photophosphorylation). Of particular relevance, photosynthesis requires large amounts of proteins (notably Rubisco) and proteins are synthesized by phosphorus-rich ribosomes (Wang et al. 2008). As a result, channelling metabolic flux to lipid biosynthesis by the means of phosphate starvation may have a severe impact on photosynthesis.

There is apparently a dilemma in the BE strategy, i.e., the very reason that stimulates lipid accumulation in cells may result in severely impeded cell growth and photosynthesis and hence lowered overall lipid productivity. This dilemma could likely be solved by employing metabolic engineering approaches aiming at enhancing the metabolic flux into lipid biosynthesis without applying the aforementioned "artificial" physiological stresses.

5.4. Lipid Production Enhancement: The Genetic Engineering (GE) Approaches

Although biotechnological processes based on transgenic microalgae are still in their infancy, researchers and companies are considering the potential of microalgae as green cell-factories to produce value-added metabolites and heterologous proteins for pharmaceutical applications (Leon-Banares et al. 2004). The commercial application of algael transgenics is beginning to be realized and algael biotechnology companies are being established. It was predicted that microalgae, due to the numerous advantages they present, could offer a powerful tool for the production of commercial molecules in a near future (Cadoret et al. 2008).

The fast growing interests in the use of transgenic microalgae for industrial applications is powered by the rapid developments in microalgael biotechnology. Complete genome sequences from the red algae *Cyanidioschyzon merolae* (Nozaki et al. 2007), the diatoms *Thalassiosira pseudonana* (Armbrust et al. 2004) and *Phaeodactylum tricornutum* (Bowler et al. 2008) and the unicellular green algae *Ostreococcus tauri* (Derelle et al. 2006) have been completed. Nuclear transformation of various microalgael species is now a routine, chloroplast transformation has been achieved for green, red, and euglenoid algae, and further success in organelle transformation is likely as the number of sequenced plastid, mitochondrial, and nucleomorph genomes continues to grow (Walker et al. 2005a). Various genetic transformation systems have been developed in green algae such as *Chlamydomonas reinhardtti* and *Volvox carteri* (Walker et al. 2005a).

The fast developments of microalgael biotechnology permit the isolation and use of key genes for genetic transformation. Of particular relevance, acetyl-CoA carboxylase (ACC) was first isolated from the microalgae *Cyclotella cryptica* in 1990 by (Roessler 1990) and then successfully transformed by Dunahay et al. (Dunahay et al. 1996; Dunahay et al. 1995; Sheehan et al. 1998) into the diatoms *Cyclotella cryptica* and *Navicula saprophila*. The ACC gene, *acc*1, was overexpressed with the enzyme activity enhanced to two to three-folds. These experiments demonstrated that ACC could be transformed efficiently into microalgae although no significant increase of lipid accumulation was observed in the transgenic diatoms (Dunahay et al. 1996; Dunahay et al. 1995). It also suggests that overexpression of ACC enzyme alone might not be sufficient to enhance the whole lipid biosynthesis pathway (Sheehan et al. 1998).

Even though there is no success story with respect to lipid overproduction of microalgae using the GE approach up to now, a solid understanding towards the global TAG biosynthesis pathway, which is generally accepted to be identical throughout all species except the differences in the location of reactions and the structure of some key enzymes, has been established. Extensive studies have also been carried out regarding the enhancement of lipid production using the GE approach in different species. These results provide a valuable background for future studies with microalgae.

5.4.1. Overexpression of TAG Biosynthesis Pathway Enzymes

Numerous studies have been carried using the GE strategy to enhance the lipid accumulation in different species. Some of these studies have been summarized in Table 5.1 and will be discussed briefly in this section.

5.4.1.1. Acetyl-Coa Carboxylase (ACC)

Since the successful demonstration by Page et al. (1994) that ACC exerts a strong control on the metabolic flux of fatty acid synthesis in plants, this enzyme has been overexpressed in different species for enhanced lipid production. For instance, the cytosolic ACC from *Arabidopsis* was overexpressed in *Brassica napus* (rapeseed) plastid with a one- to twofold increase of plastid ACC activity (Roesler et al. 1997). However, the fatty acid content of the recombinant was only 6% higher than the control, suggesting that a "secondary bottleneck", i.e., another limiting step, in the fatty acid synthesis pathway might have emerged as a result of the removal of the primary bottleneck. Davis et al. (2000) cloned the four ACC genes, accA, accB, accC and accD of *E. coli* BL21 and overexpressed them in the same strain. ACC subunits were produced in equimolar quantities. This caused an increase of the intracellular malonyl-CoA pool as a result of the enhanced ACC enzymatic activity. A 6-fold increase in the rate of fatty acid synthesis was observed, confirming that the ACC catalyzed committing step was indeed the rate-limiting step for fatty acid biosynthesis in this strain. However, the lack of lipid production enhancement seemed to suggest again that a secondary limiting-step after fatty acid formation prevented the efficient conversion of fatty acids to lipids in *E. coli*. As mentioned previously, ACC was also isolated from microalgae (Roessler 1990) and successfully overexpressed in the diatoms *Cyclotella cryptica* and *Navicula saprophila*. Not surprisingly, no significant increase of lipid accumulation was observed in the transgenic diatoms (Dunahay et al. 1996; Dunahay et al. 1995).

Table 5.1. Lipid synthesis enhancement genes (enzymes)

Gene (enzyme)	Source-species	Receiver-species	Note	Refs
accA, accB, accC, accD (ACC), tesA (thioesterase I)	*E, coli* (BL21) (bacteria)	*E. coli* (BL21) (bacteria)	6x fatty acid synthesis	(Davis et al. 2000)
Acc1 (cytosolic ACC)	*Arabidopsis* (plant)	*Brassica napus* (plant)	1-2x plastid ACC + 6% fatty acid content	(Roesler et al. 1997)
Acc1 (ACC)	*Arabidopsis* (plant)	*Solanum tuberosum* (plant)	5x TAG content	(Klaus et al. 2004)
Acc1 (ACC)	*Cyclotella cryptica* (algae)	*Cyclotella cryptica* (algae)	2-3x ACC activity, no change in lipid content	(Dunahay et al. 1995) (Dunahay et al. 1996)
Acc1 (ACC)	*Cyclotella cryptica* (algae)	*Navicula saprophila* (algae)	2-3x ACC activity, no change in lipid content	(Dunahay et al. 1995) (Dunahay et al. 1996)
fabF (KAS II)	*E. coli* (bacteria)	*E. coli* (bacteria)	Toxic (CoA pool from 0.5 to 40% in malonyl-CoA)	(Subrahmanyam and Cronan Jr 1998)
fabH (KAS III)	*E. coli* (bacteria)	*Brassica napus* (plant)	Stress, arrest of the cell growth	(Verwoert et al. 1995)
KAS III	*Spinacia oleracea* (plant)	*Nicotiana tabacum* (plant)	16 :0 accumulation Lower oil content	(Dehesh et al. 2001)
KAS III	*Spinacia oleracea* (plant)	*Arabidopsis* (plant)	16 :0 accumulation Lower oil content	(Dehesh et al. 2001)
KAS III	*Spinacia oleracea* (plant)	*Brassica napus* (plant)	16 :0 accumulation Lower oil content	(Dehesh et al. 2001)
LPAT	*Saccharomyces cerevisiae* (yeast)	*Brassica napus* (plant)	6x oil content	(Zou et al. 1997)
are1 and are2 (DGAT)	*Arabidopsis thaliana* (plant)	Yeast	3-9x TAG content	(Bouvier-Nave et al. 2000)
are1 and are2 (DGAT)	*Arabidopsis thaliana* (plant)	*Nicotiana Tabacum* (plant)	7x TAG content	(Bouvier-Nave et al. 2000)
DGAT	*Arabidopsis* (plant)	*Arabidopsis* (plant)	+ 10-70% oil content	(Jako et al. 2001)
acs (ACS)	*E. coli* (MG1655) (bacteria)	*E. coli* (MG1655) (bacteria)	9x ACS activity, increased acetate assimilation	(Lin et al. 2006)
malEMt and malEMc (ME)	*Mortierella alpina* & *Mucor circinelloides* (fungi)	*Mucor circinelloides* (fungi)	2.5x lipid accumulation	(Zhang et al. 2007)
ACL	Rat	Tobacco	+ 16% lipid content	(Rangasamy and Ratledge 2000)
Antisens PEP gene	*Agrobacterium tumefaciens*	*Brassica napus*	+ 6.4-18% oil content	(Chen et al. 1999)

* thioesterase I

Sheehan et al. (1998) suggested that overexpression of ACC enzyme alone may not be sufficient to enhance the whole lipid biosynthesis pathway in diatoms. This conclusion seems to be generally true for most species because enhanced lipid accumulation was rarely reported even though significant enhancement of relevant enzymes and/or intermediate products such as fatty acids was commonly observed. This is probably due to either or both of the following two reasons 1) the committing step catalyzed by ACC is not the rate-limiting step in a particular species; and 2) a secondary rate-limiting step, i.e. the "secondary bottleneck", emerged when ACC was overexpressed. Nevertheless, Klaus et al. (2004) achieved an increase in fatty acid synthesis and a more than 5-fold increase in the amount of TAG in *Solanum tuberosum* (potato) by overexpressing the ACC from *Arabidopsis* in the amyloplasts of potato tubers.

5.4.1.2. Fatty Acid Synthetase (FAS)

Trials in overexpressing the KAS subunit of FAS in *E. coli* were carried out to facilitate the C2 concatenation. However, this manipulation was found extremely toxic for the cell (Subrahmanyam and Cronan Jr 1998). In another trial, an *E. coli* KAS III was overexpressed in the rapeseed (Verwoert et al. 1995), which caused a major change in the fatty acid composition profile with the increase of short-chain fatty acids (14:0) and a decrease of 18:1 fatty acids. This modification caused a response to stress, which significantly affected the growth of the plant cells. Similarly, KAS III from spinach *Spinacia oleracea* was overexpressed by Dehesh et al. (2001) in tobacco *Nicotiana tabacum*, cress *Arabidopsis* and rapeseed, resulted in a reduction of the rate of lipid synthesis and an accumulation of 16:0 fatty acids.

It seems that the subunits of FAS are challenging targets for metabolic engineering for fatty acid metabolism enhancement, probably due to the fact that FAS is a multi-enzymatic complex containing subunits whose activities depend on one another. The difficulties experienced with the heterologous expression of multi-enzymatic complexes such as FAS were also likely due to the differences in multipoint controls among different species.

5.4.1.3. Lysophosphatidate Acyl-Transferase (LPAT)

Transformation of rapeseed with a putative *sn*-2-acyl-transferase gene from the yeast *Saccharomyces cerevisiae* was carried out by Zou et al. (1997), leading to overexpression of seed lysophosphatidate acyl-transferase (LPAT) activity. This enzyme is involved in TAG formation and its overexpression led to increases from 8 to 48% seed oil content on the seed dry weight basis. However, it was cautioned that the steady-state level of diacylglycerol could be perturbed by an increase of LPAT activity in developing seeds.

5.4.1.4. Acyl-CoA:Diacylglycerol Acyl-Transferase (DGAT)

Acyl-CoA:diacylglycerol acyl-transferase (DGAT) catalyzes, as discussed previously, the last step of TAG formation to form triacylgycerol from diacylglycerol and fatty acyl CoA. Two full-length cDNAs of *Arabidopsis* encoding proteins of 520 and 532 amino acids, respectively, were confirmed to encode acyl CoA:diacylglycerol acyl-transferases. Transformations of yeast and tobacco, respectively, with the *Arabidopsis* DGAT were performed. A 200 to 600-fold increase of DGAT activity in the transformed yeast was observed, which led to a three- to nine-fold increase of TAGs accumulation. In the transformed tobacco, TAG content increased to sevenfold higher than that of a control plant. In addition, lipid droplets formation occurred in the cytoplasm of young growing leaf cells as a result of this transformation (Bouvier-Nave et al. 2000). DGAT gene has also been overexpressed in the plant *Arabidopsis* and it was shown that the oil content was enhanced in correlation with the DGAT activity, which increased by 10 to 70% (Jako et al. 2001).

The success with DGAT could be explained by the fact that the substrate of DGAT, diacylglycerol, could be allocated to either phospholipid biosynthesis or TAG formation. Overexpression of DGAT would commit more diacylglycerol to TAG formation rather than phospholipid formation. In fact, studies with plants have revealed that increasing the rate of TAG synthesis by overexpressing DGAT also stimulated the formation of fatty acid (Galili and Hofgen 2002). All these results seem to suggest that the reaction catalyzed by DGAT is an important rate-limiting step in lipid biosynthesis. However, no reports regarding the overexpression of this enzyme in microalgae were located.

5.4.2. Overexpression of Enzymes Relevant to Lipid Biosynthesis

A few enzymes that are not directly involved in lipid metabolism have also been demonstrated to influence the rate of lipid accumulation by increasing the pool of essential metabolites for lipid biosynthesis. The following are a few examples.

5.4.2.1. The Acetyl-Coa Synthase (ACS)

ACS catalyzes the conversion of acetate into acetyl-CoA. It was observed that when growing a bacterial strain on acetate, overexpression of ACS could increase the rate of fatty acid synthesis. For instance, it was observed by Lin et al. (2006) that, by over-expressing the *acs* gene in *E. coli,* the ACS activity was increased by 9-fold, leading to a significant increase of the assimilation of acetate from the medium, which can contribute to lipid biosynthesis. This concept was also shown to be applicable to the secreted acetate during bacterial growth (Brown et al. 1977).

5.4.2.2. Overexpression of Malic Enzyme (ME)

The effect of ME, which catalyzes the conversion of malate into pyruvate and simultaneously reduces a $NADP^+$ molecule into NADPH, was studied in filamentous fungi in correlation with lipid accumulation (Wynn et al. 1999). It was observed that the enhanced energy (NADPH) supply as a result of ME overexpression was utilized by the enzymes involved in TAG synthesis and led to enhanced lipid production. It was observed that the enhanced activity of ME led to the increase of the cytosolic NADPH pool (i.e., the reducing equivalent that reflects the cellular energy state), making available more reducing power for lipogenic enzymes such as ACC, FAS and ATP:citrate lyase (ACL). It was proposed that a metabolon formation between FAS and ME could take place to create a channeling of the NADPH formed by ME toward the FAS active sites.

A similar strategy, i.e., to allow lipogenesis to occur without energy restriction by overexpressing ME so that the lipid accumulation becomes maximal, was investigated recently by Zhang et al. (2007) with *Mucor circinelloides*. The genes coding for ME from *M. circinelloides* (malEMt) and from *Mortierella alpine* (malEMc), respectively, were overexpressed in *M. circinelloides*. Two and three-fold increases in ME activity were observed for the transgenic malEMt and malEMc strains, respectively. In both cases, the ME activity increase was associated with a faster lipid accumulation. The amount of synthesized lipids was 2.5- and 2.4-fold higher for the transgenic malEMt and malEMc strains, respectively.

5.4.2.3. The ATP:Citrate Lyase (ACL)

ACL catalyzes the conversion of citrate into acetyl-CoA and oxaloacetate, and thus represents a source of acetyl-CoA for fatty acid biosynthesis. It has been well established that ACL is a key enzyme in lipid accumulation regulation in mammals, oleaginous yeast and fungi. It was also demonstrated that heterologous ACL can be imported into the plastids of plants. Rangasamy and Ratledge (2000) did an interesting experiment in which a gene encoding a fusion protein of the rat liver ACL with the leader peptide for the small subunit of ribulose bisphosphate carboxylase was constructed and introduced into the genome of tobacco. This was sufficient to transport the heterogonous enzyme into the plastids. In vitro assays of ACL in isolated plastids showed that the enzyme was active and synthesized acetyl-CoA. Overexpression of the rat ACL gene led to a fourfold increase in the total ACL activity; this increased the amount of fatty acids by 16% but did not cause any major change in the fatty acid profile.

5.4.3. Blocking Competing Pathways

From the metabolic engineering point of view, blocking off competing pathways may also enhance the metabolic flux being channeled to TAG biosynthesis.

β-oxidation is the principal metabolic pathway responsible for the degradation of fatty acids in eukaryotes (Shen and Burger 2008). By doing so, it consumes fatty acids, the precursors of TAG formation. It is therefore possible to enhance TAG production by blocking this pathway. Cao et al.(2006) demonstrated that using an indirect method, i.e., inhibiting the acetyl-CoA transportation system required for coupling the β-oxidization in peroxisome and the TCA cycle in mitochondria but not any enzyme of the β-oxidation, is capable of inhibiting the β-oxidization of *Candida tropicali*. Dicarboxylic acids (DCAs) can be obtained by oxidizing alkanes by *C. tropicalis*. However, DCAs may be degraded to acetyl-CoA by β-oxidation, resulting in a limited DCA yield. In *C. tropicali*, acetyl-CoA can be transported into the mitochondrion for the TCA cycle by carnitine acetyl-transferase (CAT), by which the energy generation and β-oxidation are connected. It was shown that the reduction of the specific activity of CAT in recombinant cells by about 50% resulted in a 21.0% increase of the DCA concentration, and a 12% increase of the molar conversion of alkane. However, recombinants with no detectable CAT activity could not grow on alkane. These results indicate that partial inhibition of β-oxidation can facilitate DCA production. However, complete blocking of the transportation process would be harmful for energy supply. Picataggio et al. (1992) blocked β-oxidation in *Candida tropicalis* by knocking out the genes encoding for acyl-CoA oxydase. It was observed that, not surprisingly, the growth of the cells was adversely affected.

Phospholipid biosynthesis is another competitive pathway to TAG formation because it competes against TGA biosynthesis for a common substrate, phosphatidate. If phosphatidate is converted into CDP-diacylglycerol instead of diacylglycerol, it enters the phospholipids synthetic pathway (Coleman and Lee 2004). As mentioned previously, overexpression of the enzyme DGAT has the effect of channeling phosphatidate to TGA accumulation. On the other hand, it was shown that inhibition of phospholipid synthesis caused the formation of abnormally long fatty acids, due to supplementary elongation cycles (Jiang and Cronan Jr 1994).

The third competitive pathway is the conversion of phosphoenolpyruvate to oxaloacetate, which is catalyzed by the phosphoenolpyruvate carboxylase (PEPC). TAG biosynthesis also requires phosphoenolpyruvate (which converts successively to pyruvate, acetyl-CoA, malonyl-CoA and then fatty acids) (Song et al. 2008). By expressing antisense PEPC in *B. Napus*, Chen et al. (1999) achieved a 6.4 to 18% increase in oil content, suggesting that reduced PEPC activity enhanced the lipid accumulation. Significantly enhanced lipid contents were also obtained with transgenic soybean lines harbouring anti-PEP gene (Sugimoto et al. 1989; Zhao et al. 2005). In microalgae, preliminary results also indicate that PEPC plays a role in the regulation of fatty acid accumulation and reduced PEPC activity by antisense expression was correlated with an increase of the lipid content in *Synechococcus sp*., a cyanobacterium (Song et al. 2008).

5.4.4. The Multi-Gene Approach

The multi-gene approach, i.e., overexpressing more than one key enzymes in the TAG pathway to enhance lipid biosynthesis, was suggested by a few researchers (Roesler et al. 1997; Verwoert et al. 1995). However, literature on the feasibility of this strategy is scarce, probably due to the difficulties in manipulating multiple genes.

In summary, extensive studies have established a solid understanding of the lipid metabolism in different species. Based on the knowledge, numerous trials have been carried out to investigate the feasibility of manipulating the genes of key enzymes relevant to lipid synthesis to enhance lipid production of different species. They can be broadly classified into four different approaches: 1) overexpressing rate-limiting enzymes of the TAG biosynthesis pathway; 2) overexpressing enzymes that enhance the TAG pathway; 3) partially blocking competing pathways; and 4) the multi-gene transgenic approach. It seems that DGAT and ME are the most likely enzymes that might lead to enhanced lipid production when overexpressed in plants. However, no report was found regarding the overexpression of either of these two enzymes in microalgae. While completely blocking a competing pathway seems to be harmful to cell growth, preliminary successes have been achieved with the partial repression of CAT and

PEPC. Of particular interest, reducing PEPC activity by expressing antisense gene was observed to be beneficial for lipid production in microalgae.

5.5. Lipid Production Enhancement: The Transcription Factor Engineering (TFE) Approach

It was recently suggested that the regulation of metabolic pathways must be studied in the context of the whole cell rather than at the single pathway level. The use of regulatory factors such as transcription factors (TFs) to control the abundance or activity of multiple enzymes relevant to the production of desired products has provoked widespread interests (Capell and Christou 2004). This approach is referred to as TFE and can be more precisely described as a novel technology employing the overexpression of TFs that up- or down-regulate the pathway(s) being involved in the formation of target metabolites for the overproduction of them.

TFs are proteins that regulate DNA transcription by recognizing specific DNA sequences and establishing protein-DNA and protein-protein interactions. They have been classified into more than 50 families according to their conserved structure and their DNA binding domains. They can interact with the transcription machinery such as DNA polymerase and so to activate it to enhance the rate of transcription of a particular group of genes (Grotewold 2008). They can also act as repressors or make subtle down-regulation changes in a metabolite production without repressing it totally. Quite frequently, a combination of TFs may regulate a single metabolic pathway (Santos and Stephanopoulos 2008).

In contrast to the GE approach that targets a single gene, the TF approach affects a large number of genes involving multiple metabolic pathways, resulting in an integrated up- or down-regulation of these pathways simultaneously (Grotewold 2008; Santos and Stephanopoulos 2008). The emerging of "secondary bottlenecks", which is one of the major concerns of the GE approaches, is therefore less likely. This emerging metabolic engineering approach has been demonstrated to be able to improve the production of valuable metabolites (see Table 5.2) and represents an attractive alternative that is likely to bring out the breakthroughs in producing metabolically engineered microalgael strains for cost-effective TAG production.

Although TFE in microalgae is still in its embryo, numerous TFs have been shown to be able to stimulate the overproduction of valuable metabolites in different species and various TFs for the regulation of lipid synthesis in animals, plants and microorganisms have been identified. These results may provide valuable hints to TFE for enhanced microalgael lipid production.

Table 5.2. Transcription factors enhanced production of high-value products

TF	Source-species (taxonomy)	Receiver-species (taxonomy)	Effectiveness	Refs
Artificial zinc fingers	Artificial	Tobacco (plant)	High level activation of a β-glucuronidase gene stable, inheritable, non-toxic	(Segal et al. 2003)
Zinc fingers	Human	CHO cells	Twofold increase of IgG antibody production	(Reik et al. 2007)
MYB & bHLH	*Arabidopsis thaliana* (plant)	*Arabidopsis thaliana* (plant)	Strongly enhanced flavonoïd biosynthesis	(Vom Endt et al. 2002)
ORCA2	*Catharanthus roseus* (plant)	*Catharanthus roseus* (plant)	Induction of genes leading to TIA biosynthesis	(Vom Endt et al. 2002)

5.5.1. Enhanced Metabolite Production by TFE

5.5.1.1. Zinc-Finger Protein Transcription Factors for Enhanced Pharmaceutical Proteins

Zinc-finger protein transcription factors (ZFP TF), which typically contain many fingers linked in a tandem fashion, are some of the most extensively studied DNA-binding proteins. The zinc finger domain enables different proteins to interact with or bind DNA, RNA, or other proteins, and is present in the proteomes of a variety of different organisms. There are many types of zinc finger proteins, which are classified according to the number and order of their Cys and His residues that bind the Zinc ion. Among these, the C2H2-type zinc finger proteins, which have 176 members in *Arabidopsis thaliana* alone, constitute one of the largest families of TFs in plants. They are mostly species-specific and contain a conserved QALGGH sequence within their zinc finger domain. Recent functional characterization studies of different C2H2 proteins in *Arabidopsis* suggest that many of these proteins function as part of a large regulatory network that senses and responds to different environmental stimuli (Ciftci-Yilmaz and Mittler 2008). Based on the understanding to the structural and functional features of naturally occurring zinc finger proteins, several design strategies have been proposed for the creation of artificial zinc-finger proteins for applications in gene regulation and gene therapy (Negi et al. 2008; Segal et al. 2003; Stege et al. 2002). Enhanced production of a therapeutic protein was achieved by overexpressing a ZFP TF that binds a DNA sequence within the promoter of a therapeutic protein from mammalian production cell lines (Reik et al. 2007). This ZFP TF enabled more than 100% increase in protein yield from CHO cells. Expression vectors engineered to carry up to 10 ZFP binding sites further enhanced ZFP-mediated increases in protein production up to 500%.

5.5.1.2. MYB and Bhlh Transcription Factors for Enhanced Production of Flaronoids in Plants

MYB and bHLH transcription factors have been studied in plants such as *Arabidopsis thaliana* and have been demonstrated to regulate the biosynthesis of flavonoids, more precisely anthocyanin and seed coat tannin (Vom Endt et al. 2002). When genes R and C1, which encoded a bHLH and a MYB protein, respectively, were ectopically expressed in normally unpigmented cell lines, accumulation of anthocyanin was observed. This was the consequence of a coordinate response to the TFs in the form of a global expression of the structural genes (Vom Endt et al. 2002). Similarly, overexpression of MYB in *Arabidopsis* caused a significant enhancement of the flavonoid biosynthesis (Vom Endt et al. 2002).

5.5.1.3. ORCA2 Protein for Enhanced Alkaloid Production in Plants

Plant alkaloids are a source of many novel natural products such as pharmaceuticals. Several TFs involved in the regulation of plant alkaloid biosynthesis genes have been isolated and studied. Among them is ORCA2 protein, the TF stimulates terpenoid indole alkaloid (TIA) biosynthesis. It was shown that by overexpressing ORCA2, multiple genes of the TIA biosynthesis pathway were overexpressed, leading to an increase of TIA formation (Vom Endt et al. 2002).

5.5.2. TFE for Enhanced Microalgael Lipid Production: How Far are We from There?

To be able to implement the TFE strategy for lipid production in microalgae, TFs from algae have to be identified. A few TFs have been identified as responsible for the regulation of lipid biosynthesis in animals and plants. For instance, sterol regulatory element binding protein (SREBP) have been established as the master regulators of lipid homeostasis in mammals (Eberle et al. 2004; Espenshade 2006; Espenshade and Hughes 2007; Goldstein et al. 2006; Hitoshi 2005; Horton 2002; McPherson and Gauthier 2004; Porstmann et al. 2005; Todd et al. 2006; Yang et al. 2000). In plants, it was demonstrated that SebHLH protein, a member of the bHLH family TFs, might play a key role in the transcriptional regulation of genes related to storage lipid biosynthesis and accumulation during seed development (Kamisaka et al. 2007). It was also reported that soybean Dof-type (DNA binding with one finger) TF genes were involved in the regulation of the lipid content in soybean seeds. Among the 28 Dof-type TF genes in soybean plants, which displayed diverse patterns of expression in various organs and exhibited different abilities for transcriptional activation and DNA binding, two genes, GmDof4 and GmDof11, were found to increase the content of total fatty acids and lipids in transgenic *Arabidopsis* seeds by upregulating genes that are associated with the biosynthesis of fatty acids (Wang et al. 2007a). The Dof-type TF family sequences were also identified from a variety of representative organisms from different taxonomic groups: the unicellular green algae *Chlamydomonas reinhardtii*, the moss *Physcomitrella patens*, the club moss *Selaginella moellendorffii*, the gymnosperm *Pinus taeda*, the dicotyledoneous *Arabidopsis thaliana* and the monocotyledoneous angiosperms *Oryza sativa* and *Hordeum vulgare* (Moreno-Risueno et al. 2007). It is worth noting that SREBP proteins, the master regulator of mammalian lipid homeostasis, are also found in plants and microorganisms with high conservation of sequences (Bengoechea-Alonso and Ericsson 2007; Espenshade and Hughes 2007; Todd et al. 2006). Nevertheless, their regulatory functions are very different from those in mammals. For instance, in fission yeast *Schizosaccharomyces pombe*, the SREBP analog, which is called Sre1p, was found to be a principal activator of anaerobic gene expression, upregulating genes required for nonrespiratory oxygen consumption, among many other up-regulated genes, while down-regulating a large number of other genes. It was observed that oxygen-requiring biosynthetic pathways for ergosterol, heme, sphingolipid, and ubiquinone were the primary targets of Sre1p, which acted directly at target gene promoters (Todd et al. 2006).

TFs can be classified into a pyramidal hierarchy, where the high-level TFs influence the low-level ones (Grotewold 2008). The predictability of the TFs is variable and depends on their level in the hierarchy. A high-level TF has important impacts on other TFs and in consequence regulates a broad range of genes. It follows that, quite logically, predicting the outcome of these high-level TFs on global gene expression is still troublesome with our limited understanding at present. On the other hand, low-level TFs are generally less conserved and are hence much more difficult to be employed in inter-species metabolic engineering manipulations. It seems to be clear from the previous discussion that the TFs regulating lipid biosynthesis are low-level TFs as different species have different TFs for lipid regulation. There is so far not much information regarding the lipid regulation TFs of microalgae.

At least 147 putative TFs and 87 putative transcription regulators (TRs) (proteins that assist TF functions) have been identified in the green algae *C. reinhardtii* up to 2008. However, only the biological functions of a small number of them have been determined (Riano-Pachon et al. 2008). As aforementioned, no literature regarding the TFs responsible for the regulation of lipid biosynthesis in microalgae was found. The most important task at present is therefore to identify lipid regulating TFs of microalgae should we plan to exploit the TFE strategy for enhanced lipid production. Fortunately, various technologies for the identification, purification, and characterization of TFs have been developed, providing a solid foundation for future studies in this direction.

A common strategy used for the identification of TFs involves comparing the transcriptomics and proteomics of target microalgae under controlled conditions that allow and prohibits the formation of metabolites of interest, respectively. For instance, Egan et al. (2002) identified a ToxR-like transcription regulator (WmpR) that controls the expression of fouling inhibitors in Pseudoalteromonas tunicate by analysing the gene sequence of a transposon mutant deficient in antifouling activities and comparing the proteomics of the wildtype and the mutant strains using 2-D PAGE. Then, more precise analyses unveiling the regulator functions of WmpR were carried by Stelzer et al. (2006) using the combination of proteomic analysis (2D-PAGE) and transcriptomic studies (RNA arbitrarily primed PCR). Recently, transcriptomic studies using microarrays were employed by Nguyen et al. (2008) to identify factors regulating the hydrogen production of C. reinhardtii. In this study, microarray analyses were used to obtain global expression profiles of mRNA abundance in the green algae C. reinhardtii at different time points before the onset and during the course of sulfur-depleted hydrogen production. These studies were followed by real-time quantitative reverse transcription-PCR and protein analyses. Among the more than twofold differentially expressed genes, 10 were classified as having a putative role in transcription or translation. Four of these were upregulated during the hydrogen production phase, including the genes encoding pre-mRNA processing factor 3 (PRP3), the eukaryotic initiation factor 4A-10 (eIF4A-10), splicing factor 3a subunit 3 (SP3a3), and the chloroplast 30S ribosomal protein L11 (rps11).

A number of different techniques have been developed for purification of TFs, which is an important step prior to studies on the structure and functions of TFs (Jiang et al. 2007; Maouche and Cohen-Kaminsky 1997; Schulman and Setzer 2002; Yang 1998). In principle, these methods are all based on the ability of TFs to recognize and interact with specific DNA sequences present in the promoters of eukaryotic genes. The purification of a TF begins with the preparation of nuclear extracts from appropriate cells or tissues (Gorski et al. 1986), which is then subjected to a series of pre-treatment procedures (Yang 1998) if necessary, followed by DNA affinity chromatography (Moxley and Jarrett 2005).

Various technologies have been employed for characterizing the structure of TFs and their interaction with DNA sequences. These technologies include electrophoretic mobility shift assay (EMSA) (Hattori et al. 2007; Hickman and Harwood 2008; Wang et al. 2003; Yang 1998), the DNase I protection (footprinting) assay (Wang et al. 2007b; Yang 1998) and the Southwestern blotting (Wang et al. 2003; Yang 1998). NMR spectroscopy (Bagby et al. 1998; Yamasaki et al. 2008) and X-ray crystallography (Burley and Kamada 2002; Yamasaki et al. 2008) are also commonly used in combination with other methods to study TF structure and TF-DNA interaction.

A comprehensive knowledge of the TF binding site (TFBS) is important for the understanding of TF regulatory functions. Techniques for identifying TFBS, including experimental techniques and computational approaches, can be found in a few recent reviews (Efromovich et al. 2008; Elnitski et al. 2006; Hannenhalli 2008; Marinescu et al. 2005; Merkulova et al. 2007; Mukhopadhyay et al. 2008).

In summary, there is little doubt that TFE method is a very promising technology that is likely to bring about the necessary breakthrough enabling cost-effective microalgael oil production. However, study on the TFE for enhanced microalgael lipid production is scarce and the most important task is to identify TFs regulating microalgael lipid biosynthesis. To this end, various technologies are available for the identification, purification, and characterization of TFs and TF functions.

5.6. Conclusion

There are three promising strategies that could potentially be employed for enhancing lipid production of microalgae, the BE strategy, the GE strategy, and the TFE strategy. Firstly, the BE strategy, which relies on applying physiological stresses such as nutrient-depletion to channel metabolic flux to lipid biosynthesis, is the most mature and most widely employed among the three at present. Secondly, the GE and the TFE strategies are more promising in a long term perspective. Although there is a lack of success stories in lipid overproduction using transformed microalgael strains, the knowledge obtained in studies on lipid pathways and genetic transformed organisms for enhanced lipid synthesis among other species suggests that DGAT and ME are the most promising targets for gene transformation. Downregulation of PEPC gene to reduce the PEPC activity was also suggested to be beneficial for lipid production in some microalgael species. Finally, TFE is an emerging technology that has a great potential. In-depth studies on the physiological functions of microalgael TFs and identification of TFs regulating lipid pathways of different microalgael species are the essential steps for successful implementation of the TFE strategy. To this end, a number of techniques have been developed for the identification, purification, and characterization of TFs.

REFERENCES

Armbrust EV, Berges JA, Bowler C, Green BR, Martinez D, Putnam NH, Zhou S, Allen AE, Apt KE, Bechner M and others. 2004. The genome of the diatom Thalassiosira Pseudonana: Ecology, evolution, and metabolism. *Science* 306(5693):79-86.

Athenstaedt K, Daum G. 1999. Phosphatidic acid, a key intermediate in lipid metabolism. *European Journal of Biochemistry* 266(1):1-16.

Bagby S, Arrowsmith CH, Ikura M. 1998. New perceptions of transcription factor properties from NMR. *Biochemistry and Cell Biology* 76(2-3):368-378.

Bengoechea-Alonso MT, Ericsson J. 2007. SREBP in signal transduction: cholesterol metabolism and beyond. *Current Opinion in Cell Biology* 19(2):215-222.

Bouvier-Nave P, Benveniste P, Oelkers P, Sturley SL, Schaller H. 2000. Expression in yeast and tobacco of plant cDNAs encoding acyl CoA:diacylglycerol acyltransferase. *European Journal of Biochemistry* 267(1):85-96.

Bowler C, Allen AE, Badger JH, Grimwood J, Jabbari K, Kuo A, Maheswari U, Martens C, Maumus F, Otillar RP and others. 2008. The Phaeodactylum genome reveals the evolutionary history of diatom genomes. *Nature*.

Brown TDK, Jones-Mortimer MC, Kornberg HL. 1977. The enzymic interconversion of acetate and acetyl-coenzyme A in Escherichia coli. *Journal of General Microbiology* 102(2):327-336.

Budziszewski GJ, Croft KPC, Hildebrand DF. 1996. Uses of biotechnology in modifying plant lipids. *Lipids* 31(6):557-569.

Burley SK, Kamada K. 2002. Transcription factor complexes. *Current Opinion in Structural Biology* 12(2):225-230.

Cadoret JP, Bardor M, Lerouge P, Cabigliera M, Henriquez V, Carlier A. 2008. *Microalgae as cell factories producing recombinant commercial proteins. Les microalgues: Usines cellulaires productrices de molecules commerciales recombinantes* 24(4):375-382.

Cao Z, Gao H, Liu M, Jiao P. 2006. Engineering the acetyl-CoA transportation system of candida tropicalis enhances the production of dicarboxylic acid. *Biotechnology journal.* 1(1):68-74.

Capell T, Christou P. 2004. Progress in plant metabolic engineering. *Current Opinion in Biotechnology* 15(2):148-154.

Chen JQ, Lang CX, Hu ZH, Liu ZH, Huang RZ. 1999. Antisense PEP gene regulates to ratio of protein and lipid content in Brassica napus seeds. *J Agr Biotech* 7(4):316-320.

Chisti Y. 2007. Biodiesel from microalgae. *Biotechnology Advances* 25(3):294-306.

Ciftci-Yilmaz S, Mittler R. 2008. The zinc finger network of plants. *Cellular and Molecular Life Sciences* 65(7-8):1150-1160.

Coleman RA, Lee DP. 2004. Enzymes of triacylglycerol synthesis and their regulation. *Progress in Lipid Research* 43(2):134-176.

Davis MS, Solbiati J, Cronan Jr JE. 2000. Overproduction of acetyl-CoA carboxylase activity increases the rate of fatty acid biosynthesis in Escherichia coli. *Journal of Biological Chemistry* 275(37):28593-28598.

Dehesh K. 2001. How can we genetically engineer oilseed crops to produce high levels of medium-chain fatty acids? *European Journal of Lipid Science and Technology* 103(10):688-697.

Dehesh K, Tai H, Edwards P, Byrne J, Jaworski JG. 2001. Overexpression of 3-ketoacyl-acyl-carrier protein synthase IIIs in plants reduces the rate of lipid synthesis. *Plant Physiology* 125(2):1103-1114.

Derelle E, Ferraz C, Rombauts S, Rouze P, Worden AZ, Robbens S, Partensky F, Degroeve S, Echeynie S, Cooke R and others. 2006. *Genome analysis of the smallest free-living eukaryote Ostreococcus tauri unveils many unique features.* Proceedings of the National Academy of Sciences of the United States of America 103(31):11647-11652.

Dunahay TG, Jarvis EE, Dais SS, Roessler PG. 1996. Manipulation of microalgael lipid production using genetic engineering. *Applied Biochemistry and Biotechnology - Part A Enzyme Engineering and Biotechnology* 57-58:223-231.

Dunahay TG, Jarvis EE, Roessler PG. 1995. Genetic transformation of the diatoms Cyclotella cryptica and Navicula saprophila. *J. Phycol.* 31:1004-1012.

Dyer JM, Mullen RT. 2005. Development and potential of genetically engineered oilseeds. *Seed Science Research* 15(4):255-267.

Eberle D, Hegarty B, Bossard P, Ferre P, Foufelle F. 2004. SREBP transcription factors: Master regulators of lipid homeostasis. *Biochimie* 86(11):839-848.

Efromovich S, Grainger D, Bodenmiller D, Spiro S. 2008. Genome-Wide Identification of Binding Sites for the Nitric Oxide-Sensitive Transcriptional Regulator NsrR. In: Poole RK, editor. *Methods in Enzymology.* p 211-233.

Elnitski L, Jin VX, Farnham PJ, Jones SJM. 2006. Locating mammalian transcription factor binding sites: A survey of computational and experimental techniques. *Genome Research* 16(12):1455-1464.

Espenshade PJ. 2006. SREBPs: Sterol-regulated transcription factors. *Journal of Cell Science* 119(6):973-976.

Espenshade PJ, Hughes AL. 2007. Regulation of sterol synthesis in eukaryotes. *Annual review of genetics* 41:401-427.

Galili G, Hofgen R. 2002. Metabolic engineering of fatty acid biosynthesis in plants. *Metabolic Engineering* 4(1):12-21.

Goldstein JL, DeBose-Boyd RA, Brown MS. 2006. Protein sensors for membrane sterols. *Cell* 124(1):35-36.

Gorski K, Carneiro M, Schibler U. 1986. Tissue-specific in vitro transcription from the mouse albumin promoter. *Cell* 47(5):767-776.

Graham IA, Larson T, Napier JA. 2007. Rational metabolic engineering of transgenic plants for biosynthesis of omega-3 polyunsaturates. *Current Opinion in Biotechnology* 18(2):142-147.

Grotewold E. 2008. Transcription factors for predictive plant metabolic engineering: are we there yet? *Current Opinion in Biotechnology* 19(2):138-144.

Hannenhalli S. 2008. Eukaryotic transcription factor binding sites - Modeling and integrative search methods. *Bioinformatics* 24(11):1325-1331.

Hattori H, Imai H, Kirai N, Furuhama K, Sato O, Konishi K, Nakagawa Y. 2007. Identification of a responsible promoter region and a key transcription factor, CCAAT/enhancer-binding protein , for up-regulation of PHGPx in HL60 cells stimulated with TNF *Biochemical Journal* 408(2):277-286.

Hickman JW, Harwood CS. 2008. Identification of FleQ from Pseudomonas aeruginosa as a c-di-GMP-responsive transcription factor. *Molecular Microbiology* 69(2):376-389.

Hitoshi S. 2005. Lipid synthetic transcription factor, SREBP. Nippon rinsho. *Japanese journal of clinical medicine* 63(5):897-907.

Horton JD. 2002. Sterol regulatory element-binding proteins: Transcriptional activators of lipid synthesis. *Biochemical Society Transactions* 30(6):1091-1095.

Huntley ME, Redalje DG. 2007. CO2 mitigation and renewable oil from photosynthetic microbes: A new appraisal. Mitigation and Adaptation *Strategies for Global Change* 12(4):573-608.

Illman AM, Scragg AH, Shales SW. 2000. Increase in Chlorella strains calorific values when grown in low nitrogen medium. *Enzyme and Microbial Technology* 27(8):631-635.

Ivy JM, Beremand PD, Thomas TL. 1998. Strategies for modifying fatty acid composition in transgenic plants. *Biotechnology and Genetic Engineering Reviews* 15:271-288.

Jako C, Kumar A, Wei Y, Zou J, Barton DL, Giblin EM, Covello PS, Taylor DC. 2001. Seed-specific over-expression of an arabidopsis cDNA encoding a diacylglycerol acyltransferase enhances seed oil content and seed weight. *Plant Physiology* 126(2):861-874.

Jiang P, Cronan Jr JE. 1994. Inhibition of fatty acid synthesis in Escherichia coli in the absence of phospholipid synthesis and release of inhibition by thioesterase action. *Journal of Bacteriology* 176(10):2814-2821.

Jiang S, Nagore L, Jiang D, Moxley RA, Jarrett HW. 2007. Purification of transcription factors and identification with mass spectrometry. *Current Analytical Chemistry* 3(4):272-282.

Kamisaka Y, Tomita N, Kimura K, Kainou K, Uemura H. 2007. DGA1 (diacylglycerol acyltransferase gene) overexpression and leucine biosynthesis significantly increase lipid accumulation in the snf2 disruptant of Saccharomyces cerevisiae. *Biochemical Journal* 408(1):61-68.

Kawata M, Nanba M, Matsukawa R, Chihara M, Karube I. 1998. Isolation and characterization of a green algae Neochloris sp. for CO2 fixation. *Studies in Surface Science and Catalysis* 114:637-640.

Khozin-Goldberg I, Cohen Z. 2006. The effect of phosphate starvation on the lipid and fatty acid composition of the fresh water eustigmatophyte Monodus subterraneus. *Phytochemistry* 67(7):696-701.

Kim KH. 1997. Regulation of mammalian acetyl-coenzyme A carboxylase. *Annual Review of Nutrition*. p 77-99.

Klaus D, Ohlrogge JB, Neuhaus HE, Dormann P. 2004. Increased fatty acid production in potato by engineering of acetyl-CoA carboxylase. *Planta* 219(3):389-396.

Leon-Banares R, Gonzalez-Ballester D, Galvan A, Fernandez E. 2004. Transgenic microalgae as green cell-factories. *Trends in Biotechnology* 22(1):45-52.

Li SJ, Cronan Jr JE. 1993. Growth rate regulation of Escherichia coli acetyl coenzyme A carboxylase, which catalyzes the first committed step of lipid biosynthesis. *Journal of Bacteriology* 175(2):332-340.

Li Y, Horsman M, Wang B, Wu N, Lan CQ. 2008a. Effects of Nitrogen Sources on Cell Growth and Lipid Accumulation of Green Algae Neochloris oleoabundans. *Applied Microbiology and Biotechnology* 81 629-636.

Li Y, Horsman M, Wu N, Lan CQ, Dubois-Calero N. 2008b. Biofuels from Microalgae. . *Biotech Prog* 24: 815-820

Lin H, Castro NM, Bennett GN, San KY. 2006. Acetyl-CoA synthetase overexpression in Escherichia coli demonstrates more efficient acetate assimilation and lower acetate accumulation: A potential tool in metabolic engineering. *Applied Microbiology and Biotechnology* 71(6):870-874.

Liu ZY, Wang GC, Zhou BC. 2008. Effect of iron on growth and lipid accumulation in Chlorella vulgaris. *Bioresource Technology* 99(11):4717-4722.

Maouche L, Cohen-Kaminsky S. 1997. Strategies to identify target genes for transcription factors. *Strategies d'identification des genes cibles de facteurs de transcription* 3(1):77-88.

Marinescu VD, Kohane IS, Riva A. 2005. MAPPER: A search engine for the computational identification of putative transcription factor binding sites in multiple genomes. *BMC Bioinformatics* 6.

McPherson R, Gauthier A. 2004. Molecular regulation of SREBP function: The Insig SCAP connection and isoform-specific modulation of lipid synthesis. *Biochemistry and Cell Biology* 82(1):201-211.

Merkulova TI, Oshchepkov DY, Ignatieva EV, Ananko EA, Levitsky VG, Vasiliev GV, Klimova NV, Merkulov VM, Kolchanov NA. 2007. Bioinformatical and experimental approaches to investigation of transcription factor binding sites in vertebrate genes. *Biochemistry* (Moscow) 72(11):1187-1193.

Moreno-Risueno MA, Martinez M, Vicente-Carbajosa J, Carbonero P. 2007. The family of DOF transcription factors: From green unicellular algae to vascular plants. *Molecular Genetics and Genomics* 277(4):379-390.

Moxley RA, Jarrett HW. 2005. Oligonucleotide trapping method for transcription factor purification systematic optimization using electrophoretic mobility shift assay. *Journal of Chromatography* A 1070(1-2):23-34.

Mukhopadhyay A, Deplancke B, Walhout AJM, Tissenbaum HA. 2008. Chromatin immunoprecipitation (ChIP) coupled to detection by quantitative real-time PCR to study transcription factor binding to DNA in Caenorhabditis elegans. *Nature Protocols* 3(4):698-709.

Murphy DJ. 2001. The biogenesis and functions of lipid bodies in animals, plants and microorganisms. *Progress in Lipid Research* 40(5):325-438.

Napier JA. 2007. The production of unusual fatty acids in transgenic plants. *Annual Review of Plant Biology*. p 295-319.

Napier JA, Beaudoin F, Michaelson LV, Sayanova O. 2004. The production of long chain polyunsaturated fatty acids in transgenic plants by reverse-engineering. *Biochimie* 86(11):785-792.

Negi S, Imanishi M, Matsumoto M, Sugiura Y. 2008. New redesigned zinc-finger proteins: design strategy and its application. *Chemistry* (Weinheim an der Bergstrasse, Germany) 14(11):3236-3249.

Nguyen AV, Thomas-Hall SR, Malnoeì^ A, Timmins M, Mussgnug JH, Rupprecht J, Kruse O, Hankamer B, Schenk PM. 2008. Transcriptome for photobiological hydrogen production induced by sulfur deprivation in the green algae Chlamydomonas reinhardtii. *Eukaryotic Cell* 7(11):1965-1979.

Nozaki H, Takano H, Misumi O, Terasawa K, Matsuzaki M, Maruyama S, Nishida K, Yagisawa F, Yoshida Y, Fujiwara T and others. 2007. A 100%-complete sequence reveals unusually simple genomic features in the hot-spring red algae Cyanidioschyzon merolae. *BMC Biology* 5.

Oelkers P, Cromley D, Padamsee M, Billheimer JT, Sturley SL. 2002. The DGA1 gene determines a second triglyceride synthetic pathway in yeast. *Journal of Biological Chemistry* 277(11):8877-8881.

Opsahl Ferstad HG, Rudi H, Ruyter B, Refstie S. 2003. Biotechnological approaches to modify rapeseed oil composition for applications in aquaculture. *Plant Science* 165(2):349-357.

Page RA, Okada S, Harwood JL. 1994. Acetyl-CoA carboxylase exerts strong flux control over lipid synthesis in plants. Biochimica et Biophysica Acta - *Lipids and Lipid Metabolism* 1210(3):369-372.

Picataggio S, Rohrer T, Deanda K, Lanning D, Reynolds R, Mielenz J, Eirich LD. 1992. Metabolic engineering of Candida tropicalis for the production of long-chain dicarboxylic acids. *Bio/Technology* 10(8):894-898.

Porstmann T, Griffiths B, Chung YL, Delpuech O, Griffiths JR, Downward J, Schulze A. 2005. SREBP PKB/Akt induces transcription of enzymes involved in cholesterol and fatty acid biosynthesis via activation of SREBP. *Oncogene* 24(43):6465-6481.

Rajakumari S, Grillitsch K, Daum G. 2008. Synthesis and turnover of non-polar lipids in yeast. *Progress in Lipid Research* 47(3):157-171.

Rangasamy D, Ratledge C. 2000. Genetic enhancement of fatty acid synthesis by targeting rat liver ATP:citrate lyase into plastids of tobacco. *Plant Physiology* 122(4):1231-1238.

Ratledge C. 2002. Regulation of lipid accumulation in oleaginous micro-organisms. *Biochemical Society Transactions* 30(6):1047-1050.

Reik A, Zhou Y, Collingwood TN, Warfe L, Bartsevich V, Kong Y, Henning KA, Fallentine BK, Zhang L, Zhong X and others. 2007. Enhanced protein production by engineered zinc finger proteins. *Biotechnology and Bioengineering* 97(5):1180-1189.

Riano-Pachon DM, Correa LGG, Trejos-Espinosa R, Mueller-Roeber B. 2008. Green transcription factors: A Chlamydomonas overview. *Genetics* 179(1):31-39.

Rock CO, Jackowski S. 2002. Forty years of bacterial fatty acid synthesis. *Biochemical and Biophysical Research Communications* 292(5):1155-1166.

Roesler K, Shintani D, Savage L, Boddupalli S, Ohlrogge J. 1997. Targeting of the Arabidopsis homomeric acetyl-coenzyme A carboxylase to plastids of rapeseeds. *Plant Physiology* 113(1):75-81.

Roessler PG. 1990. Purification and characterization of acetyl-CoA carboxylase from the diatom Cyclotella cryptica. *Plant Physiology* 92(1):73-78.

Sandager L, Gustavsson MH, Stahl U, Dahlqvist A, Wiberg E, Banas A, Lenman M, Ronne H, Stymne S. 2002. Storage lipid synthesis is non-essential in yeast. *Journal of Biological Chemistry* 277(8):6478-6482.

Santos CNS, Stephanopoulos G. 2008. Combinatorial engineering of microbes for optimizing cellular phenotype. *Current Opinion in Chemical Biology* 12(2):168-176.

Sasaki Y, Nagano Y. 2004. Plant acetyl-CoA carboxylase: Structure, biosynthesis, regulation, and gene manipulation for plant breeding. *Bioscience, Biotechnology and Biochemistry* 68(6):1175-1184.

Schenk P, Thomas-Hall S, Stephens E, Marx U, Mussgnug J, Posten C, Kruse O, Hankamer B. 2008. Second Generation Biofuels: High-Efficiency Microalgae for Biodiesel Production. *BioEnergy Research* 1(1):20-43.

Schulman DB, Setzer DR. 2002. Identification and characterization of transcription factor IIIA from Schizosaccharomyces pombe. *Nucleic Acids Research* 30(13):2772-2781.

Scragg AH, Illman AM, Carden A, Shales SW. 2002. Growth of microalgae with increased calorific values in a tubular bioreactor. *Biomass and Bioenergy* 23(1):67-73.

Segal DJ, Stege JT, Barbas Iii CF. 2003. Zinc fingers and a green thumb: Manipulating gene expression in plants. *Current Opinion in Plant Biology* 6(2):163-168.

Sendl A, Schliack M, Loser R, Stanislaus F, Wagner H. 1992. Inhibition of cholesterol synthesis in vitro by extracts and isolated compounds prepared from garlic and wild garlic. *Atherosclerosis* 94(1):79-85.

Sheehan TDJ, Benemann J, Roessler P. 1998. *A look back at the U.S. Department of energy's aquatic species program-biodiesel from algae*. NREL/TP-580-24190.

Shen YQ, Burger G. 2008. Plasticity of a key metabolic pathway in fungi. *Functional and Integrative Genomics*:1-7.

Song D, Fu J, Shi D. 2008. Exploitation of Oil-bearing Microalgae for Biodiesel. *Chinese Journal of Biotechnology* 24(3):341-348.

Stege JT, Guan X, Ho T, Beachy RN, Barbas Iii CF. 2002. Controlling gene expression in plants using synthetic zinc finger transcription factors. *Plant Journal* 32(6):1077-1086.

Stelzer S, Egan S, Larsen MR, Bartlett DH, Kjelleberg S. 2006. Unravelling the role of the ToxR-like transcriptional regulator WmpR in the marine antifouling bacterium Pseudoalteromonas tunicata. *Microbiology* 152(5):1385-1394.

Stoll C, Luhs W, Zarhloul MK, Friedt W. 2005. Genetic modification of saturated fatty acids in oilseed rape (Brassica napus). *European Journal of Lipid Science and Technology* 107(4):244-248.

Subrahmanyam S, Cronan Jr JE. 1998. Overproduction of a functional fatty acid biosynthetic enzyme blocks fatty acid synthesis in Escherichia coli. *Journal of Bacteriology* 180(17):4596-4602.

Sugimoto T, Tanaka K, Monma M, Kawamura Y, Saio K. 1989. Phosphoenolpyruvate carboxylase level in soybean seed highly correlates to its contents of protein and lipid. *Agric. Biol. Chem.* 53:885-887.

Takagi M, Karseno, Yoshida T. 2006. Effect of salt concentration on intracellular accumulation of lipids and triacylglyceride in marine microalgae Dunaliella cells. *Journal of Bioscience and Bioengineering* 101(3):223-226.

Tehlivets O, Scheuringer K, Kohlwein SD. 2007. Fatty acid synthesis and elongation in yeast. Biochimica et Biophysica Acta - *Molecular and Cell Biology of Lipids* 1771(3):255-270.

Todd BL, Stewart EV, Burg JS, Hughes AL, Espenshade PJ. 2006. Sterol regulatory element binding protein is a principal regulator of anaerobic gene expression in fission yeast. *Molecular and Cellular Biology* 26(7):2817-2831.

Tornabene TG, Holzer G, Lien S, Burris N. 1983. Lipid composition of the nitrogen starved green algae *Neochloris oleoabundans*. *Enzyme and Microbial Technology* 5(6):435-440.

Verwoert II, Van Der Linden KH, Walsh MC, Nijkamp HJ, Stuitje AR. 1995. Modification of Brassica napus seed oil by expression of the Escherichia coli fabH gene, encoding 3-ketoacyl-acyl carrier protein synthase III. *Plant Molecular Biology* 27(5):875-886.

Vom Endt D, Kijne JW, Memelink J. 2002. Transcription factors controlling plant secondary metabolism: What regulates the regulators? *Phytochemistry* 61(2):107-114.

Walker TL, Collet C, Purton S. 2005a. Algael transgenics in the genomic era. *Journal of Phycology* 41(6):1077-1093.

Walker TL, Purton S, Becker DK, Collet C. 2005b. Microalgae as bioreactors. *Plant Cell Reports* 24(11):629-641.

Wang B, Li Y, Wu N, Lan CQ. 2008. CO2 bio-mitigation using microalgae. *Applied Microbiology and Biotechnology* 79:707-718.

Wang C, Yeung F, Liu PC, Attar RM, Geng J, Chung LWK, Gottardis M, Kao C. 2003. Identification of a novel transcription factor, GAGATA-binding protein, involved in androgen-mediated expression of prostate-specific antigen. *Journal of Biological Chemistry* 278(34):32423-32430.

Wang HW, Zhang B, Hao YJ, Huang J, Tian AG, Liao Y, Zhang JS, Chen SY. 2007a. The soybean Dof-type transcription factor genes, GmDof4 and GmDof11, enhance lipid content in the seeds of transgenic Arabidopsis plants. *Plant Journal* 52(4):716-729.

Wang X, Kikuchi T, Rikihisa Y. 2007b. Proteomic identification of a novel Anaplasma phagocytophilum DNA binding protein that regulates a putative transcription factor. *Journal of Bacteriology* 189(13):4880-4886.

World Bank. 2008. Focus B, *Biofuels: Promise and Risks*.

Wynn JP, Hamid ABA, Ratledge C. 1999. The role of malic enzyme in the regulation of lipid accumulation in filamentous fungi. *Microbiology* 145(8):1911-1917.

Yamaberi K, Takagi M, Yoshida T. 1998. Nitrogen depletion for intracellular triglyceride accumulation to enhance liquefaction yield of marine microalgael cells into a fuel oil. *Journal of Marine Biotechnology* 6(1):44-48.

Yamasaki K, Kigawa T, Inoue M, Watanabe S, Tateno M, Seki M, Shinozaki K, Yokoyama S. 2008. Structures and evolutionary origins of plant-specific transcription factor DNA-binding domains. *Plant Physiology and Biochemistry* 46(3):394-401.

Yang T, Goldstein JL, Brown MS. 2000. Overexpression of membrane domain of SCAP prevents sterols from inhibiting SCAP-SREBP exit from endoplasmic reticulum. *Journal of Biological Chemistry* 275(38):29881-29886.

Yang VW. 1998. Eukaryotic transcription factors: Identification, characterization and functions. *Journal of Nutrition* 128(11):2045-2051.

Zhang Y, Adams IP, Ratledge C. 2007. Malic enzyme: The controlling activity for lipid production? Overexpression of malic enzyme in Mucor circinelloides leads to a 2.5-fold increase in lipid accumulation. *Microbiology* 153(7):2013-2025.

Zhao G, Chen J, Yin A. 2005. Transgenic soybean lines harbouring anti-PEP gene express super-high oil content. *Mol Plant Bre* 3:792-796.

Zou J, Katavic V, Giblin EM, Barton DL, MacKenzie SL, Keller WA, Hu X, Taylor DC. 1997. Modification of seed oil content and acyl composition in the Brassicaceae by expression of a yeast sn-2 acyltransferase gene. *Plant Cell* 9(6):909-923.

Chapter 6

OPTIMIZING LIPID PRODUCTION OF THE GREEN ALGAE *NEOCHLORIS OLEOABUNDANS* USING BOX-BEHNKEN EXPERIMENTAL DESIGN IN COMBINATION WITH FACTOR GROUPING

6.1. INTRODUCTION

As discussed in Chapter 4, biodiesel is one of the most promising renewable biofuels that have achieved remarkable success worldwide. It is predictable that massive global demand on renewable energy will continue to drive the rapid growth of biodiesel production in an unprecedented scale (Courchesne et al.) and oil-rich microalgae have been demonstrated to be a promising candidate (Chisti 2007, Courchesne et al. , Li et al. 2008a).

Several microalgael species have been investigated for lipid production. It was reported by Reitan et al. (1994) that *Chlorella emersonii*, *Chlorella minutissima*, and *Chlorella vulgaris* could accumulate lipids of up to 63%, 56%, and 57.9%, respectively under nitrogen starvation conditions. Takagi et al. (1998) reported *Dunaliella*, a marine algae, accumulated a high intracellular lipid content of 67% when the initial NaCl concentration in medium was 1.0 M (seawater equivalent to 0.5 M NaCl). *Neochloris Oleoabundans* was reported to be able to accumulate 35-54% lipids of its cell dry weight and triglycerides comprised 80% of the total lipids (Tornabene et al. 1983). Of particular relevance, as discussed in Chapter 6, *N. oleoabundans* has been demonstrated to be capable of producing lipids at a productivity of 0.133 g l^{-1} day^{-1} at 5 mM with a lipid cell content of

0.34 g/g and a biomass productivity of 0.40 g l^{-1} day^{-1}(Li et al. 2008b). Nevertheless, the primary challenge impeding commercial microalgael production remains the relatively high production costs of microalgael cultivation, which is expected to be overcome by extensive studies including medium optimization. Systematic studies on the effects of different nutritional and operational factors on cell growth and lipid biosynthesis of oil-rich microalgael strains are therefore of both scientific and practical relevance. Understanding of these effects is essential for formulation of optimal medium to enhance oil production and reduce production costs.

Most studies regarding the cultivation of microalgae have been carried out using single factor searching, which is based on investigating the effects of a single nutrient on the physiology of microalgae while fixing the composition of other nutrients. The single factor searching approach, although capable of providing valuable information, is in general tedious and of low efficiency. It is especially so for the optimization of microalgael media considering the large number of nutrients contained in a microalgael medium. It is therefore of value to develop a more efficient approach for microalgael medium optimization.

Statistic experimental design such as BBD, which allows the study of multiple factors and their interactions using workable number of factors, is getting popularity for medium optimization for different microbial systems (Annadurai et al. 1999, Bae and Shoda 2005, Ramnani and Gupta 2004). However, information regarding the application of this approach with microalgae is scarce in the literature. Furthermore, the extraordinarily large number of nutrients involved in the medium of some microalgael species makes it difficult to handle even with statistical experimental designs. In this study, we employed an integrated approach to systematically study the effects of different nutrients on cell growth and lipid production of *N. oleoabundans*. This approach involves the grouping of different nutrients according to their effects on the physiology of microalgae, which were then treated as individual single factors in pre-studies and statistical investigations. The Box-Behnken experimental design (BBD) was used for statistical studies and five macroelements, i.e., nitrogen, phosphorus, magnesium, iron, and sulphur, were selected for studies using lipid productivity as the response. It was demonstrated that the grouping strategy, when combined with carefully designed BBD experiments, allowed high-efficiency systematic studies on the effects and co-effects of a large number of medium components.

6.2. Materials and Methods

6.2.1. Microalgael Strain and Cultivation Conditions

Green algae *N. oleoabundans* was purchased from the UTEX Culture collection of Algae (UTEX, Texas). The algae culture was adapted to cultivation temperature of 28±2°C before use in the optimization studies.

Experiments were conducted in 500 ml cylinder flasks with 400ml working volume. Agitation was provided by a magnetic stirrer. The culture was bubbled continuously with filtered air enriched with 5% CO_2 at a flowrate of 0.75 vvm. The cultivation temperature was maintained at 28±2°C. Continuous illumination at 6000 lux was provided by 6×100w fluorescent lights, measured with a quantum meter (model QLS, Biospherical Instruments Inc., CA, USA). The initial biomass concentration for all cultures was controlled in the range of 0.2-0.22 (OD_{600nm}).

6.2.2. Media

Modified Bristol Medium, which was comprised of Bristol medium (Tam and Wong, 1996) plus 1 ml A_5 solution per 1000 ml medium, was used as the basic medium for this study. The Bristol medium consist of the following components (g/L): $NaNO_3$ (0.25), K_2HPO_4 (0.075), KH_2PO_4 (0.175) and $MgSO_4$ (0.037) , $FeCl_3$ (0.003), $CaCl_2.2H_2O$ (0.025), NaCl (0.025), and the A_5 solution is comprised of the following components (mg/L): EDTA-Fe (1,642.3), H_3BO_3 (2,860), $MnCl_2.4H_2O$ (1,810), $ZnSO_4.7H_2O$ (220), $CuSO_4.5H_2O$ (79), and $(NH_4)_6MO_7O_{24}.4H_2O$ (39). In the optimization studies, medium composition varied as indicated in the text to investigate the effects of different nutrients on cell growth and lipid production.

6.2.3. Analytical Methods

Biomass concentration was determined turbidometrically at 600 nm using a spectrophotometer (GENESYS 10UV, Thermo Electron Co., USA). Samples were diluted to give an OD_{600} reading between 0.2 – 0.4 if applicable. The OD_{600} reading was then multiplied with a pre-determined conversion factor of 0.4 to obtain the dry cell weight (DCW g/l).

The chlorophyll *a* content of cells was determined as follows: 2 ml of algael suspension was centrifuged at 12000 rpm for 12 min to collect algael cells, which were washed with distilled water for three times and then re-suspended in 1 ml methanol and stored overnight at a temperature of 4°C. The suspension was then centrifuged at 12000 rpm for 12 min and the absorbance of the supernatant was measured at wavelengths 665 nm and 650 nm, respectively. The chlorophyll *a* concentration in the extract (supernatant) was calculated using the following equation (Aslan and Kapdan, 2006):

$$\text{Chlorophyll } a \text{ (mg/L)} = (16.5 \times A_{665}) - (8.3\times A_{650}) \quad (1)$$

The lipid content of cells was determined using Soxhlet extraction. Algael biomass was first dried and grounded thoroughly in a laboratory oven at 90°C. The dried powders were then placed in the thimble of a Soxhlet extractor and the lipid was extracted using 60 ml of ethyl ether for 12 h. Then, the residue was dried and extracted again using approximately 60 ml ethyl ether. After extraction, ethyl ether was evaporated and lipid determined gravimetrically. The lipid content was calculated by the following equation:

$$\text{Lipid content (\%)} = (\text{weight of lipid/ weight of dried algael cells}) \times 100 \quad (2)$$

6.3. Factor Grouping and Experimental Design

6.3.1. The Pre-Studies

This study was carried out in two stages, pre-studies and BBD optimization. In pre-studies, all the 13 different salts used in the medium of *N. oleoabundans* cultivation (Li et al. 2008b) were divided into two groups according to their concentrations in the medium: microelements and macronutrients. Their effects on cell growth were investigated by treating these two groups of nutrients as two independent factors using the single factor searching approach.

6.3.2. The BBD Experimental Design

BBD design and RSM analysis were employed to elucidate the effects of five important macronutrient salts on cell growth and lipid production of *N.*

oleoabundans. Factor grouping was used to reduce the number of factors to increase the efficiency of the optimization studies.

6.3.2.1. Factor Selection and Grouping

Nitrogen is the constituent element of a large number of essential cell materials (Lourenco et al. 2002) including proteins, nucleic acids, cell wall, pigments, etc. It has been well established that nitrogen-sufficiency promotes high growth rates at the price of lowered oil accumulation in cells, whereas nitrogen-deficiency reduced growth rate and resulted in high oil cell content. In fact, nitrogen is the most commonly used limiting-factor for the stimulation of lipid accumulation in microalgae (Courchesne et al. 2009, Li et al. 2008b, Wang et al. 2008). It is therefore selected as a single factor in the BBD experiments.

Phosphorus is another key macroelement that has significant relevance to virtually every metabolic process of cell growth (Agren 2004). It is a constituent element of DNA, RNA, adenosine triphosphate (ATP), cell membrane materials, etc. ATP is essential to the cellular processes related to photosynthesis (e.g., photophosphorylation) and other energetic metabolisms (Gauthier and Turpin 1997, Martinez et al. 1999), which are critical for cell growth and lipid production of microalgae. Phosphorus is preferentially assimilated as inorganic phosphates in the form of $H_2PO_4^-$ and HPO_4^{2-}, which could serve as the buffering agents for pH control as well. Phosphate, in the form of NaH_2PO_4/Na_2HPO_4 pair, was therefore selected as another factor for optimization.

Magnesium, sulphur and iron are the other three important macronutrients that have significant effects on microalgael physiology and metabolism. Of particular relevance, magnesium is a constituent element of chlorophyll molecule, which is the key device of photosynthesis. Both iron and magnesium could regulate the uptake of other essential elements and enhance the production of lipids (Roden and Zachara 1996). Furthermore, Magnesium is required for nitrogenase activity using a creatine phosphate/kinase/ATP generating system as one of its roles in cell metabolism and iron is involved in electron flow from H_2O to nicotinamide adenine dinucleotide phosphate (Raven et al. 1999). Sulfur is an essential component of cysteine and methionine. In the absence of sulfur, protein biosynthesis is impeded and the photosynthetic system PSII repair cycle is blocked (Zhang et al. 2002). These three elements are supplied by two salts, magnesium sulfate and ferrous chloride. These two salts (i.e., the three elements) are required in small quantities in comparison to that of N and P and were grouped as a single factor to reduce the number of factors.

Accordingly, concentrations of $NaNO_3$, the KH_2PO_4/K_2HPO_4 pair at a fixed ratio of 17.5:7.5 w/w, and the $MgSO_4/FeCl_3$ pair at a fixed ratio of 37:3 w/w were chosen to be the three factors of BBD experiments and designated as X_1, X_2 and X_3, respectively (Table 6.1). The ratios of the KH_2PO_4/K_2HPO_4 pair and the $MgSO_4/FeCl_3$ pair followed that in the basic medium, MBM. The numbers shown in Table 6.1 were the total amounts of corresponding salt pairs.

Table 6.1. Coded and real concentration values of variables

Independent Variables(g/L)	Coded symbols	Levels -1 0 1
$NaNO_3$	X_1	0.25 0.375 0.50
K_2HPO_4-KH_2PO_4 (0.43:1)	X_2	0.25 0.5 0.75
$MgSO_4$ –$FeCl_3$ (12:1)	X_3	0.04 0.08 0.12

6.3.2.2. Levels

Since the pre-study results showed that the optimal values of most macronutrients locate between one to three times of that of the MBM medium (see discussion in later sections), one, two, , and three times of the concentration of a salt or a salt pair in MBM were set as levels -, 0 and + for the BBD experiments, respectively, however, for nitrogen source, one, 1.5, and twofold of nitrate was set as levels -, 0 and + due to our previous investigation of nitrogen effect of lipid accumulation (Li et al, 2008). (Table 6.1) The actual design of 15 experiments is given in Table 6.2. Computation was carried out using multiple regression analysis using the least squares method.

6.3.2.3. Result Analysis and Data Fitting

The BBD experimental results were fitted with the following second-order polynomial equation using a multiple regression technique:

$$Y = \beta_0 + \sum \beta_i x_i + \sum \beta_{ii} x_i^2 + \sum \beta_{ij} x_i x_j \quad j \tag{3}$$

Where Y is the predicted response (lipid yield in this study, mg/l), β_0, β_i, β_{ii}, β_{ij} are constant coefficients, and x_i, x_j (i = 1-3; j =1-3, $i \neq j$) represent the independent variables (medium composition) in the form of coded values. The quality of fit of the second-order model equation was expressed by the coefficient of determination, R^2, and its statistical significance was determined by an F-test. Canonical analysis was also carried out to predict the shape of the curve generated by the model. 3D response surface analysis was conducted by keeping one independent variable at constant level, changing the other two independent variables. The computer software used was jmp (version 6.0, James M. Pleasants Co., Inc., USA)

Table 6.2. Experimental design and results of dry weight (DW) and lipid yield

Series	$NaNO_3$	K_2HPO_4-KH_2PO_4 (0.43:1)	$MgSO_4$ –$FeCl_3$ (12:1)	DCW (g/L) Experiment	Predicted	Lipid (mg/L) P Experiment	Predicted
1	0	0	0	2.21	2.25	499.2	496.0
2	0	-	-	1.71	1.57	122.7	90.0
3	-	-	0	1.86	1.88	354.7	349.1
4	-	+	0	1.59	1.48	407.0	316.2
5	0	0	0	2.31	2.25	488.1	495.9
6	+	0	+	2.38	2.25	202.6	158.5
7	-	0	-	1.12	1.23	97.8	142.0
8	-	0	+	1.65	1.57	289.9	321.0
9	+	+	0	1.79	1.69	173.4	163.8
10	+	0	-	0.94	1.03	36.1	4.3
11	0	0	0	2.23	2.25	500.67	496.0
12	0	-	+	1.95	2.00	420	398.3
13	0	+	+	1.79	1.88	217	222.3
14	+	-	0	2.12	2.17	133.6	195.9
15	0	+	-	0.95	0.79	189.2	197.9

6.4. Results and Discussion

6.4.1. Effects of Microelements and Macronutrients on Cell Growth: Pre-Study Results

In pre-studies, two groups of nutrients, i.e., microelements with the ratios fixed as that of the A5 solution and macronutrients with ratios fixed as that of BM medium, were treated as two independent factors. Effects of microelements and macronutrients on cell growth of *N. oleoabundans* were investigated by varying their “compositions” in cultivation media.

To investigate the effects of microelements, a volume of 1, 2, 3, 4, or 5 ml of A5 solution was added in 1000 ml medium to formulate media containing one to fivefold of microelements that in MBM. As shown in Fig. 1, varying the composition of microelements in the tested range did not result in significant difference in cell growth of *N. oleoabundans*, indicating that microelements were not significant factors to *N. oleoabundans* growth within the range tested in this study.

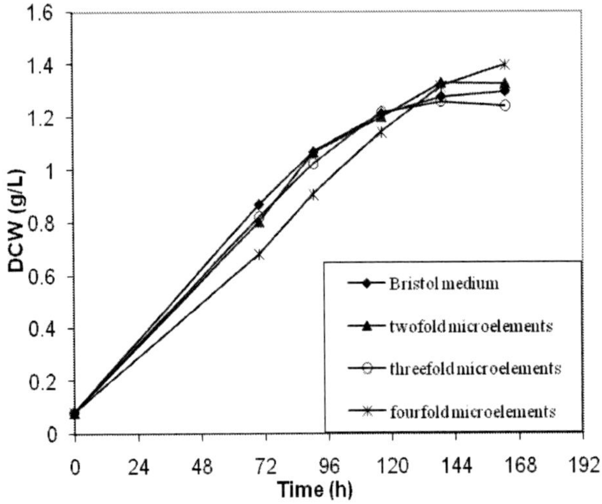

Figure 6.1. Cell growth of *N. oleoabundans* in media containing different strengths of microelements.

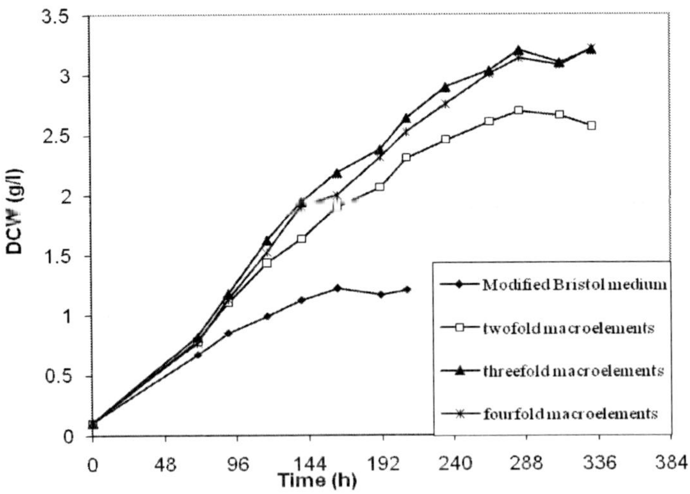

Figure 6.2. Cell growth of *N. oleoabundans* in media with varied macroelements concentrations.

Similarly, the composition of macronutrients in medium was varied between one to fivefold as that in MBM to investigate the effects of macronutrient on cell growth of *N. oleoabundans*. As shown in Fig. 2, notably improved cell growth was obtained when the macronutrient concentration increased up to threefold. However, further increase from threefold to fourfold did not produce substantial difference in cell growth. Maximum biomass in the threefold medium was above 3.0 g/L, compared favourably to that obtained with the original basic medium, which was 1.2 g/L. This result indicates that some macronutrients were deficient in the BMB and the optimal concentrations of these limiting macronutrients most likely locate in the range of one to threefold of that in MBM. The lack of variation of cell growth in media containing threefold to fivefold of macronutrients in MBM indicates that they are not inhibitive in this range.

6.4.2. BBD Studies on the Effects of Important Macronutrients

The effects of five important macronutrients, N, P, S, Fe, and Mg, which were supplemented in the form of five different salts, were studied using the BBD design and analysed using RSM. As discussed before, these five salts were grouped into three groups, which were treated as three independent factors: $NaNO_3$, the nitrogen source; K_2HPO_4/KH_2PO_4 pair at a fixed ratio of 7.5:17.5 (w/w), the phosphorus sources; and $MgSO_4/FeCl$ pair at a fixed ratio of 37:3(w/w), the source of sulphur, iron, and magnesium. The grouping of different nutrients allowed the reduction of the number of factors from five to three, which can be investigated using 15 experiments with the BBD design. The experimental design and the results of the 15 experiments are shown in Tables 1 and 2. The results were analyzed using the Analysis of Variance (ANOVA) method and the resulting parameters calculated based on dry cell weight (DCW) and lipid yield as responses are presented in Table 6.3 and Table 6.4, respectively. Substituting the coefficients (β_{ij}) in Table 6.3 and 4 to Eq. 3 results in Eqs 4 and 5, respectively:

$$C_{\text{CDW}}(g/L) = 2.251 + 0.124X_1 - 0.225X_2 + 0.386X_3 - 0.018X_1X_2 + 0.223X_1X_3 + 0.165X_2X_3 - 0.242X_1^2 - 0.203X_2^2 - 0.486X_3^2 \quad (4)$$

$$C_{\text{Lipid}}(\text{mg/L}) = -75.518X_1 - 16.811X_2 + 83.18X_3 - 0.651X_1X_2 - 6.393X_1X_3 - 71.047X_2X_3 - 154.788X_1^2 - 84.399X_2^2 - 184.59X_3^2 + 495.98 \quad (5)$$

Eqs 4 and 5 represent the quantitative effects of individual factors and their interactions on the response, C_{DCW} or C_{lipid}. The coefficients as shown in Tables 3 and 4 are related to the effects of these factors on DCW (Table 6.3) and lipid productivity (Table 6.4). The multiple correlation coefficients (R^2) of the regression equation for DCW and lipid yield obtained from ANOVA were 0.96 and 0.95, respectively, indicating that these quadratic equations could adequately describe the relationships between the factor and the responses. The sufficiency of the models is also demonstrated by Figure 6.3, which shows good correlation between the experimental values of DCW and lipid yield and corresponding values calculated using equation 4 and equation 5, respectively.

Table 6.3. Analysis of variance (ANOVA) for dry weight from BBD design

Source	**Coefficient (*β*)**	**Standard error**	***t*-Value**	***P*-value***
Intercept	2.2507	0.086	26.16	<0.0001*
X_1	0.1237	0.0527	2.34	0.066
X_2	-0.2249	0.0576	-3.90	0.0114*
X_3	0.3865	0.0527	7.33	0.0007*
X_1X_2	-0.01759	0.079	-0.22	0.8327
X_1X_3	0.2233	0.0745	3.00	0.0302*
X_2X_3	0.1654	0.079	2.09	0.0905
X_1^2	-0.2423	0.0776	-3.12	0.0261*
X_2^2	-0.2032	0.088	-2.31	0.0689
X_3^2	-0.4856	0.0776	-6.26	0.0015*

Significant at *P*-value less than 0.05

Table 6.4. Analysis of variance (ANOVA) for Lipid yield from BBD design

Source	Coefficient (*β*)	Standard error	*t*-Value	*P*-value*
Intercept	493.98	35.3028	14.05	<0.0001*
X_1	-75.5185	21.6437	-3.49	0.0175*
X_2	-16.8108	23.6332	-0.71	0.5087
X_3	83.1804	21.6437	3.84	0.0121*
X_1X_2	-0.6512	32.4274	-0.02	0.9848
X_1X_3	-6.3925	30.5732	-0.21	0.8426
X_2X_3	-71.047	32.4274	-2.19	0.0800
X_1^2	-154.7875	31.8215	-4.86	0.0046*
X_2^2	-84.3989	36.0898	-2.34	0.0665
X_3^2	-184.59	31.8215	-5.80	0.0021*

Significant at *P*-value less than 0.05

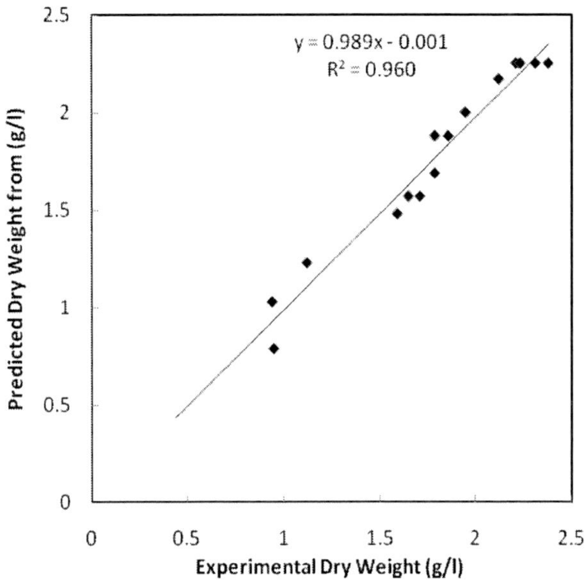

(a)

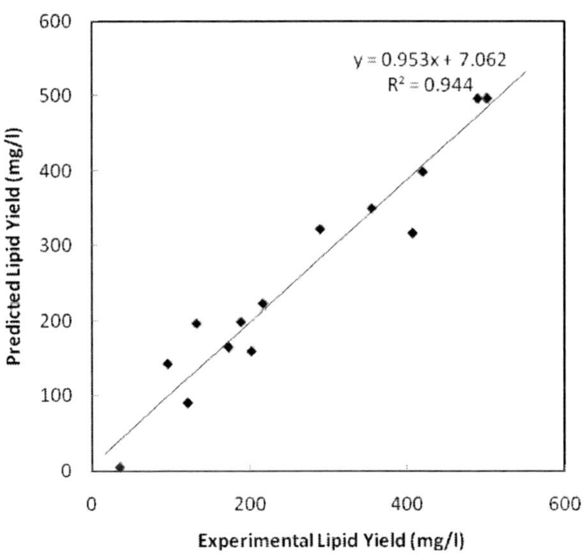

(b)

Figure 6.3. Comparison of experimental data and the predicted data: (a) dry cell weight (DCW) and (b) lipid yield.

6.4.3. Significant Factors Affecting Cell Growth and Lipid Production

A model term is considered to be significant when the P-value is less than 0.05. As shown in Table 6.3, when DCW is used as the response, K_2HPO_4/KH_2PO_4 (X_2, P = 0.0114) and $MgSO_4/FeCl_3$ (X_3, P = 0.0007) are significant factors while the effects of $NaNO_3$ (X_1) on cell growth are less than significant in the rested range. Furthermore, the interaction between X_1 and X_3 shows significant effects on cell growth of *N. oleoabundans*. This result is in accordance with the observation of Agren (2004) that the P/N ratio has significant effects on autotrophs. As shown in Tables 4, the concentration of $NaNO_3$ (X_1, P = 0.0175) and that of the $MgSO_4/FeCl_3$ pair (X_3, P = 0.0121) are significant factors for lipid production while the phosphate salt pair (X_2) showed less than significant effects.

Three-dimensional response surfaces were generated to visualize the combined effects on the lipid yield, which are shown in Fig. 4. When the effect of two factors was plotted, the other factor was set at level zero as described in Table 6.1.

Fig. 4a shows the effects of nitrate and phosphate on lipid production. Lipid yield increases with an increase of nitrate at the low concentration range. However, it decreases slightly in the high nitrate concentration range. Apparently, there is an optimal value of sodium nitrate concentration for optimized lipid production. This observation is compatible with the results of our previous studies (Li et al. 2008b). On the other hand, no significant effect on lipid production was observed by varying the concentration of the phosphate pair within the tested range.

Fig. 4b depicts the interaction between phosphate salt pair (X_2) and the magnesium sulfate and ferrous chloride pair (X_3) when the sodium nitrate concentration is fixed at level 0 (0.375 g/l). At the low range of the $MgSO_4/FeCl_3$ pair (X_3), lipid yield increases rapidly with the increase of X_3. However, lipid yield decreases slightly with the increase of X_3 after the optimal value. On the other hand, the effect of the phosphate salt pair is relatively moderate.

The effects of sodium nitrate (X_1) and $MgSO_4/FeCl$ pair (X_3) when X_2 is fixed at level zero are depicted in Fig. 6c. The optimum values of lipid correspond to approximately the coded zero level of each factor.

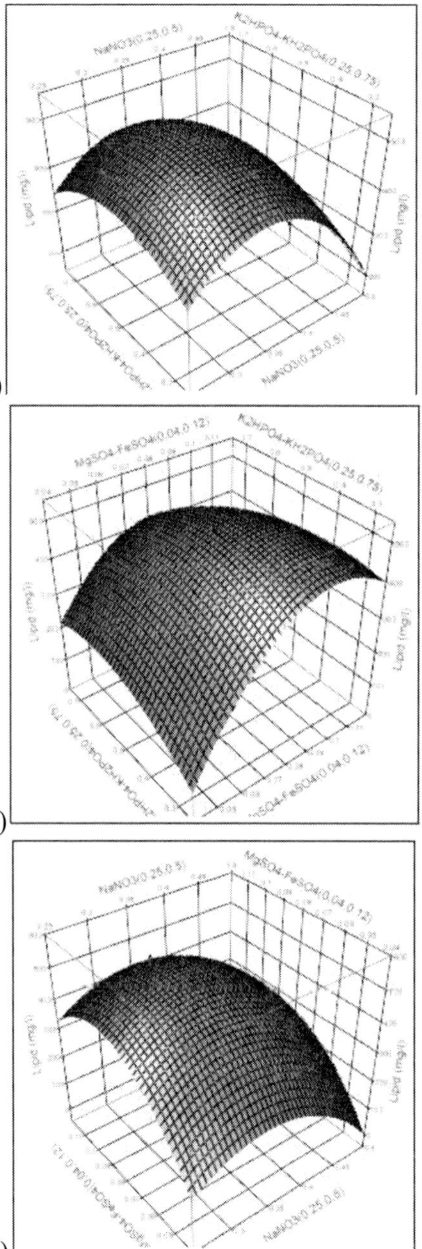

(a)

(b)

(c)

Figure 6.4. Response surface plot for lipid yield (mg/L) (a) Combined effect of $NaNO_3$ and $K_2(HPO_4)_2$-KH_2PO_4 (b) Combined effect of $K_2(HPO_4)_2$-KH_2PO_4 and $MgSO_4$ –$FeCl_3$ (c) Combined effect of $NaNO_3$ and $MgSO_4$ –$FeCl_3$.

6.4.4. Optimal Medium for Lipid Production of *N. Oleoabundans*

Among the major cell components of microalgae, i.e., proteins, carbohydrates and lipids, lipids are the most desirable from an energetic point of view(Wang et al. 2008). The biomass with a higher lipid content and lower carbohydrate and protein contents have elevated calorific values and produce higher yields of oil when processed via, for instance, biomass liquefaction (Ginzburg 1993, Illman et al. 2000, Scragg et al. 2003). Microalgael oils are also regarded as the feedstock of future for biodiesel production (Chisti 2008, Li et al. 2008a). For this reason, we optimized the medium with respect to lipid yield.

A maximum lipid yield was predicted as 603 mg/L with the optimal medium under the investigated conditions (composition in g/L): which is comprised of (g/L) $NaNO_3$ (0.35), K_2HPO_4 (0.138), KH_2PO_4 (0.322) $MgSO_4$ (0.0823), $FeCl_3$ (0.0068) and the rest the same as that in MBM. Verification of the optimal medium composition was carried out in duplicates. The time-course profiles of biomass concentration and the cell content of chlorophyll *a* are shown in Fig. 5.

As shown in Fig. 5a, cell growth was significantly improved with the optimized medium in comparison to that obtained with the basic medium. The maximum biomass density in *N. oleoabundans* cultures was two times higher than that obtained with in the basic medium. As shown in Fig. 5b, the cell contents of chlorophyll *a* of cells growing in the two media showed similar trends of change: they both increased significantly in the first day of cultivation and started to decrease from the second day or the third day of cultivation. The trend of chlorophyll *a* decreasing continued until the end of cultivation. The decrease of chlorophyll cell content was also observed in our previous studies and was explained by the fact that *N. oleoabundans* utilized chlorophyll as the internal nitrogen reservoir to support cell growth after the external nitrogen reservoir (i.e., sodium nitrate) was depleted from the culture (Li et al. 2008b).

Table 6.5. Comparison of biomass productivity and lipid productivity of *N. oleoabundans* in Modified Bristol medium (basic medium) and the optimized medium

	Basic medium	**Optimized medium**
Biomass Concentration (g/L)	1.19	2.54
Lipid Cell Content (g lipids/g DCW)	0.12	0.25
Lipid Yield (mg/l)	142.8	634.9

As summarized in Table 6.4, significant improvements were obtained with the optimized medium when comparing with the basic medium, the MBM. The maximum biomass concentration obtained with the optimized medium was 2.54 g/l, more than twofold of that obtained with MBM (1.19 g/l). The lipid yield was 635 mg/l, which was 4.5-fold of that obtained with MBM.

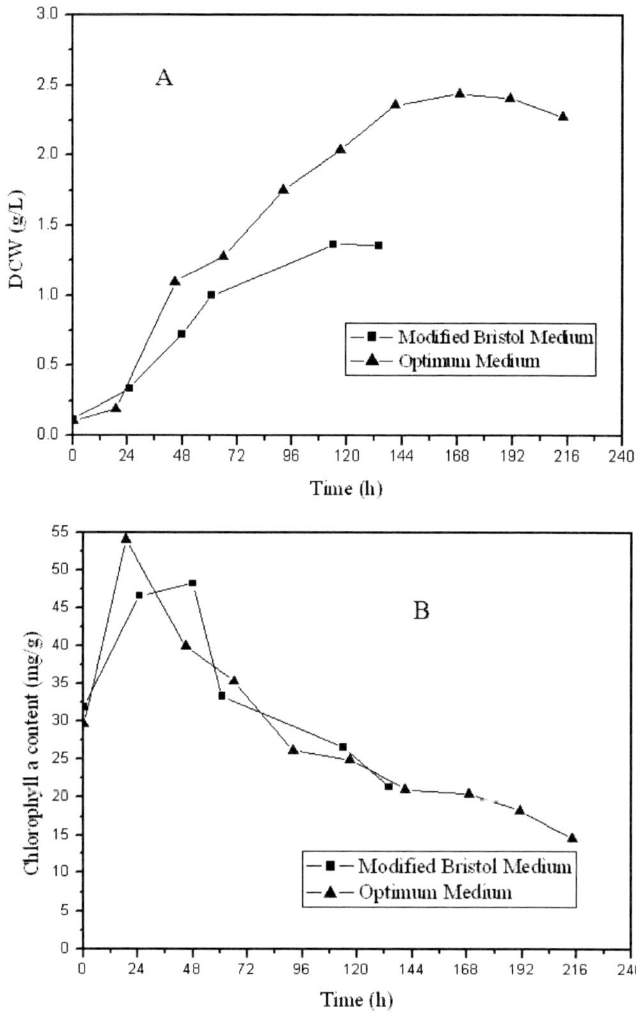

Figure 6.5. Comparison between the Modified Bristol Medium (BMB, the basic medium) and the optimized medium: (a) cell growth and (b) Chlorophyll *a* cell content.

6.5. CONCLUSIONS

In this study, the BBD with response surface analysis was successfully employed for the optimization of medium in combination with a factor grouping approach. All the nutrients were first divided according to their concentration in medium into tow groups: macronutrients and microelements, which were studied consequently as two individual factors. Several important macronutrients, including nitrogen, phosphorus, sulfate, magnesium, and iron, were then grouped into three groups, sodium nitrate as nitrogen source, phosphate salt pair as phosphorus source, and magnesium sulfate and ferrous chloride salt pair as the sources of sulphur, iron, and magnesium. The factor grouping approach made it possible to use a reasonable number of experiments to investigate the effects of a large number of nutrients, including 6 microelement-salts and 7 macronutrient-salts on the cell growth and lipid production of *N. oleoabundans*. This study demonstrates that the BBD design, when combined with careful factor selection and grouping, could be employed for efficient medium optimization.

It was determined that in the tested range, microelements have no significant effects on cell growth. Among the macronutrients, in the tested range, it was determined the phosphate salt pair has significant effects on cell growth, $NaNO_2$ has significant effects on lipid accumulation, and magnesium sulfate/ferrous chloride salt pair has significant effects on both cell growth and lipid accumulation. It was also determined that the interaction between $NaNO_3$ and the phosphate salt pair had significant effects on cell growth, even though the effect of $NaNO_3$ alone was not significant in the tested range.

An optimal medium was formulated according to the optimization results, which is comprised of (g/L) $NaNO_3$ (0.35), K_2HPO_4 (0.138), KH_2PO_4 (0.322) $MgSO_4$ (0.0823), $FeCl_3$ (0.0068) and the rest the same as that in MBM.

A biomass concentration of 2.54 g DCW/L and a lipid yield of 634.9 mg/L were obtained with the optimized medium, corresponding to 2.2-fold of cell growth and 4.5-fold of lipid yield in comparison with that obtained with the basic medium, the MBM medium.

ACKNOWLEDGMENT

Financial supports from the Natural Science and Engineering Research Council (NSERC) of Canada and the Canadian Foundation for Innovation (CFI) are gratefully acknowledged.

REFERENCES

Agren GI. 2004. The C:N:P stoichiometry of autotrophs - Theory and observations. *Ecology Letters* 7: 185-191.

Annadurai G, Balan SM, Murugesan T. 1999. Box-Behnken design in the development of optimized complex medium for phenol degradation using Pseudomonas putida (NICM 2174). *Bioprocess Engineering* 21: 415-421.

Bae S, Shoda M. 2005. Statistical optimization of culture conditions for bacterial cellulose production using Box-Behnken design. *Biotechnology and Bioengineering* 90: 20-28.

Chisti Y. 2007. Biodiesel from microalgae. Biotechnology Advances 25: 294-306.—. 2008. Biodiesel from microalgae beats bioethanol. *Trends in Biotechnology* 26: 126-131.

Courchesne NMD, Parisien A, Wang B, Lan CQ. 2009. Enhancement of lipid production using biochemical, genetic and transcription factor engineering approaches. *Journal of Biotechnology* 141: 31-41.

Gauthier DA, Turpin DH. 1997. Interactions between inorganic phosphate (P(i)) assimilation, photosynthesis and respiration in the P(i)-limited green algae Selenastrum minutum. *Plant, Cell and Environment* 20: 12-24.

Ginzburg BZ. 1993. Liquid fuel (oil) from halophilic algae: a renewable source of non- polluting energy. *Renewable Energy* 3: 249-252.

Illman AM, Scragg AH, Shales SW. 2000. Increase in Chlorella strains calorific values when grown in low nitrogen medium. *Enzyme and Microbial Technology* 27: 631-635.

Li Y, Horsman M, Wu N, Lan CQ, Dubois-Calero N. 2008a. Biofuels from microalgae. *Biotechnology Progress* 24: 815-820.

Li Y, Horsman M, Wang B, Wu N, Lan CQ. 2008b. Effects of nitrogen sources on cell growth and lipid accumulation of green algae Neochloris oleoabundans. *Applied Microbiology and Biotechnology* 81: 629-636.

Lourenco SO, Barbarino E, Mancini-Filho J, Schinke KP, Aidar E. 2002. Effects of different nitrogen sources on the growth and biochemical profile of 10 marine microalgae in batch culture: An evaluation for aquaculture. *Phycologia* 41: 158-168.

Martinez ME, Jimenez JM, El Yousfi F. 1999. Influence of phosphorus concentration and temperature on growth and phosphorus uptake by the microalgae Scenedesmus obliquus. *Bioresource Technology* 67: 233-240.

Ramnani P, Gupta R. 2004. Optimization of medium composition for keratinase production on feather by Bacillus licheniformis RGI using statistical methods involving response surface methodology. *Biotechnology and Applied Biochemistry* 40: 191-196.

Raven JA, Evans MCW, Korb RE. 1999. The role of trace metals in photosynthetic electron transport in O_2-evolving organisms. *Photosynthesis Research* 60: 111-149.

Reitan KI, Rainuzzo JR, Olsen Y. 1994. Effect of nutrient limitation on fatty acid and lipid content of marine microalgae. *Journal of Phycology* 30: 972-979.

Roden EE, Zachara JM. 1996. Microbial reduction of crystalline iron(III) oxides: Influence of oxide surface area and potential for cell growth. *Environmental Science and Technology* 30: 1618-1628.

Scragg AH, Morrison J, Shales SW. 2003. The use of a fuel containing Chlorella vulgaris in a diesel engine. *Enzyme and Microbial Technology* 33: 884-889.

Tornabene TG, Holzer G, Lien S, Burris N. 1983. Lipid Composition of the Nitrogen Starved Green Algae Neochloris Oleobundans. *Enzyme and Microbial Technology* 5: 435-440.

Wang B, Li Y, Wu N, Lan CQ. 2008. CO2 bio-mitigation using microalgae. *Applied Microbiology and Biotechnology:* 1-12.

Yamaberi K, Takagi M, Yoshida T. 1998. Nitrogen depletion for intracellular triglyceride accumulation to enhance liquefaction yield of marine microalgael cells into a fuel oil. *Journal of Marine Biotechnology* 6: 44-48.

Zhang L, Happe T, Melis A. 2002. Biochemical and morphological characterization of sulfur-deprived and H_2-producing Chlamydomonas reinhardtii (green algae). *Planta* 214: 552-561.

Chapter 7

POTENTIAL OF THE GREEN ALGAE *NEOCHLORIS OLEOABUNDANS* FOR LIPID PRODUCTION: EFFECTS OF NITROGEN SOURCES ON CELL GROWTH AND LIPID ACCUMULATION*

***Yanqun Li*[1,2]*, Bei Wang*[1]*, Nan Wu*[1]**
***and Christopher Q. Lan*[1]**

[1]University of Ottawa
[2]South China Normal University

7.1. INTRODUCTION

As discussed in previous chapters, microalgael lipids have been established as the feedstock of future for sustainable biodiesel production because microalgae have much higher growth rates and photosynthetic efficiencies than conventional terrestrial plants (Chisti 2007; Li 2008; Liu et al. 2008; Usui and Ikenouchi 1997). Nevertheless, a major obstacle for microalgael oil production is their relatively high costs, which is expected to be overcome by the technology developments. Process optimization is a very important aspect of such technology development.

---

* Most materials of this chapter were taken from an article of our group: "Effects of nitrogen sources on cell growth and lipid production of Neochloris oleoabundans", Appl Microbiol Biotechnol, 81: 629-636 (2008).

The biomass productivity, lipid cell content, and overall lipid productivity are some of the key parameters affecting the economic feasibility of algael oil for biodiesel production. While the overall lipid productivity determines the costs of the cultivation process, biomass concentration and lipid cell content affect significantly the downstream processing costs. Therefore, an ideal process should be able to produce lipid at the highest productivity with the highest lipid cell content. Unfortunately, this is not always achievable because high lipid cell contents are usually produced by cells under stress, typically nutrient limitation, which is often associated with low biomass productivity and therefore low overall lipid productivity.

A few microalgael species, including some Chlorella species (Illman et al. 2000; Liu et al. 2008; Xiong et al. 2008; Xu et al. 2006), Dunaliella species (Gordillo et al. 1998; Takagi et al. 2006), Nannochloris sp. (Takagi et al. 2000), Parietochloris incisa (Bigogno et al. 2002), N. oleoabundans (Tornabene et al. 1983), and Botryococcus braunii (Li and Qin 2005; Metzger and Largeau 2005), have been reported to have the capacity of accumulating large quantities of lipids in cells under favorable conditions. Among these species, Nannochloris and Dunaliella are marine microalgae, which are of unique advantages such as easy species control and suitable for biofuel production in coastal regions using seawater but are not suitable for freshwater cultivation. The triglycerides produced by Parietochloris incisa is rich of poly unsaturated fatty acid (Bigogno et al. 2002), making them less desirable for biodiesel production. In fact, Parietochloris incisa is considered the richest plant source of arachidonic acid (20:4ω6), containing up to 47% arachidonic acid of its total fatty acid in the stationary phase (Bigogno et al. 2002). Furthermore, the lipids produced by Botryococcus braunii (Li and Qin 2005; Metzger and Largeau 2005) are mostly large hydrocarbon molecules, which are ideal for liquid biofuel production by means of liquefaction or pyrolysis but cannot be used for biodiesel production via transesterification. On the other hand, Neochloris oleoabundans is a freshwater species that produces up to 80% triglycerides of its total lipids and most of its fatty acids are saturated fatty acid in the range of 16 – 20 carbons (Tornabene et al. 1983), ideal for biodiesel production. Quite surprisingly, however, there are no systematic studies reported in the literature regarding the application of this species for biofuel production. In this chapter, the potential of N. oleoabundans for lipids production was demonstrated using systematic studies on the effects of nitrogen source on its cell growth and lipid productivity. It is worth mentioning that N. oleoabundans has also been suggested as as excellent feed for aquaculture species such as mussel (Barnhart 2006; Gatenby et al. 2003; Gatenby et al. 1997;

Jones et al. 2005; Jones and Neves 2002), promising an alternative route of CO2 bio-sequestration using this algael species.

7.2. MATERIALS AND METHODS

Algael Strain and Cultivation Conditions

N. oleoabundans UTEX 1185, purchased from the algae culture collection at the University of Texas in Austin, was cultivated at 30±2°C with continuous illumination of 360 μmol/(m2·s1) in a modified Soil Extract (SE) Medium, which was composed of (per liter) 0.15 g K2HPO4·3H2O, 0.15 g MgSO4·7H2O, 0.05 g CaCl2·2H2O, 0.35 g KH2PO4, 0.05 g NaCl, 2.86 mg H3BO3, 1.81 mg MnCl2·4H2O, 0.22 mg ZnSO4·7H2O, 0.079 mg CuSO4·5H2O, 0.039 mg (NH4)6Mo7O24·4H2O and nitrogen source. The nitrogen source was sodium nitrate, urea or ammonium bicarbonate at the concentration indicated in the text. An enriched air stream containing 5% CO2 was first saturated with water by passing through a water bottle, and then filtered using a 0.47 micron microfiltration cartridge before being bubbled into the cultivation bottle from the bottom at a flow rate of 0.5 vvm. The cultivation bottles were 1000 ml glass columniform flasks with an inner diameter of approximately 8.5 cm and a height of approximately 18 cm. The working volume was 800 ml. The cultivation bottels were located inside a cubic box, which was equipped with 16 fluorescent lamps (tubes). Temperature inside the box was controlled by forced circulation of ambient air using two fans located on two sides of the box. Agitation was achieved by combined bubbling and magnetic stirrer.

Biomass and Lipid Measurement

Biomasses were determined routinely by measuring the optical density of samples at 600 nm (OD600). Samples were diluted by appropriate ratios to ensure that the measured OD600 values were in the range of 0.2 – 0.6 if applicable. Biomass concentration was then calculated by multiplying OD600 values with 0.4, a predetermined conversion factor converting the OD600 value to dry cell weight (DCW). The conversion factor was established by plotting OD600 versus DCW of a series of samples of different biomass concentrations. DCW of a sample was determined gravimetrically after drying the algael cells colleted from samples with centrifugation (3000 g, 10 min) and wash.

Lipid was extracted with 100 ml ethyl ether in a Soxhlet extractor from approximately 1 g dry cells (dried in an oven at 105 °C). The dry biomass was first grounded to powder and then put in the thimble of the Soxhlet extractor for extraction with reflux at 65oC for 12 hours. At the end of extraction, the extractant was distilled at 50 °C to evaporate ethyl ether, followed by drying the residue at 80 °C for 2 hours and weighting after cooling to room temperature. Lipid content was calculated by dividing the residue weight with the dry cell weight. The lipid productivity was calculated by the equation below:

$$P_{Lipid}(g/l\cdot d)=\frac{C_{Lipid}(g/g)\times DCW(g/l)}{Time(d)} \quad (1)$$

where P_{Lipid} is lipid productivity, C_{Lipid} lipid content of cells, DCW dry cell weight and Time the cultivation period in days.

Nitrate Concentration Determination

The concentrations of nitrate in media and cultivation broths were determined with HPLC method (Jovanovic et al. 2007). The measurement was performed using an HPLC system (Agilent Technologies Inc., CA, USA) loaded with an Anion IC-PAK™ column (50 mm × 4.6 mm, 10 μm particle size) (Waters, Millipore, USA). The flow rate of mobile phase was controlled at 1.2 ml/min and the UV detector set at 220 nm. The mobile phase (pH 8.5) was composed of borate buffer/gluconate concentrate, methanol, acetonitrile, and deionized water at a ratio of 2:12:12:74 (v/v/v/v). The borate buffer/gluconate concentrate was composed of 0.07 M sodium gluconate, 0.3 M H_3BO_3, 0.1 M $Na_2B_4O_7$ and 3.8 M glycerol in deionized water (100 mL). The inject volume was 50 μL.

Chlorophyll Measurement

Fresh algael cells collected in a 2 ml centrifuge tube from 2 ml cultivated broth (containing 1 – 6 mg DCW) by centrifuge were disrupted in an ultrasonic bath (FS140, output 135 W, Fisher Scientific, USA) in ice bath for 45 min and extracted with 2 ml acetone overnight. The extractant was centrifuged at 3000 g for 5 min. The contents of chlorophyll in the supernatants were determined by measuring the optical densities at 645 nm and 663 nm with a spectrophotometer and then calculated with the following equation (Becker 1994):

$$\text{Chl (mg/l)} = 8.02 \times OD_{663} + 20.21 \times OD_{645} \quad (2)$$

7.3. Results

As shown in Figure 7.1(a), ammonium bicarbonate can only support rather poor growth of *N. oleoabundans* under the investigated conditions. On the other hand, sodium nitrate was the best nitrogen source among the three tested compounds, i.e., sodium nitrate, urea, and ammonium bicarbonate. The maximum biomass concentration obtained with sodium nitrate as the nitrogen source was approximately 2.5 g/l, more than doubling that obtained with ammonium bicarbonate, which was approximately 1.1 g/l. That obtained with urea was approximately 2.1 g/l.

Microalgael biomass was harvested from culture suspension at the end of cultivation and subjected to Soxhlet extraction for lipid cell content measurement. As shown in Figure 7.1(b), the biomass obtained from the medium with sodium nitrate as nitrogen source has the highest lipid content, 38%, followed by that obtained with medium containing ammonium carbonate, 19.0%. The lipid content of cells obtained from medium using urea as the nitrogen source was approximately 17.5%, slightly lower than that obtained from the ammonium bicarbonate medium. The lipid productivity obtained with sodium nitrate was 0.133 g/l·d, which was approximately 4.7 times of that obtained with ammonium bicarbonate, 0.033 g/l·d, and 2.7 times of that obtained with urea, 0.057 g/l·d. These results suggest that sodium nitrate is the most favourable nitrogen source for both cell growth and lipid accumulation of *N. oleoabundans* under the investigated conditions.

To investigate the effect of nitrogen source concentration on cell growth and lipid accumulation of *N. oleoabundans*, modified Soil Extract Media (SE medium) containing 3, 5, 10, 15, and 20 mM sodium nitrate, respectively, were tested. The time course profiles of cell growth obtained with different sodium nitrate concentrations are shown in Figure 7.2(a). As shown in Figure 7.2(a), the cell growth of *N. oleoabundans* was improved significantly when the sodium nitrate concentration increased from 3 mM to 10 mM. However, cell growth was negatively impacted when sodium nitrate concentration further increased from 10 mM to 15 and 20 mM. As summarized in Table 7.1, both the maximum biomass concentration and biomass productivity reached the highest values at 10 mM, which were 3.15 g/l and 0.63 g/l·d, respectively.

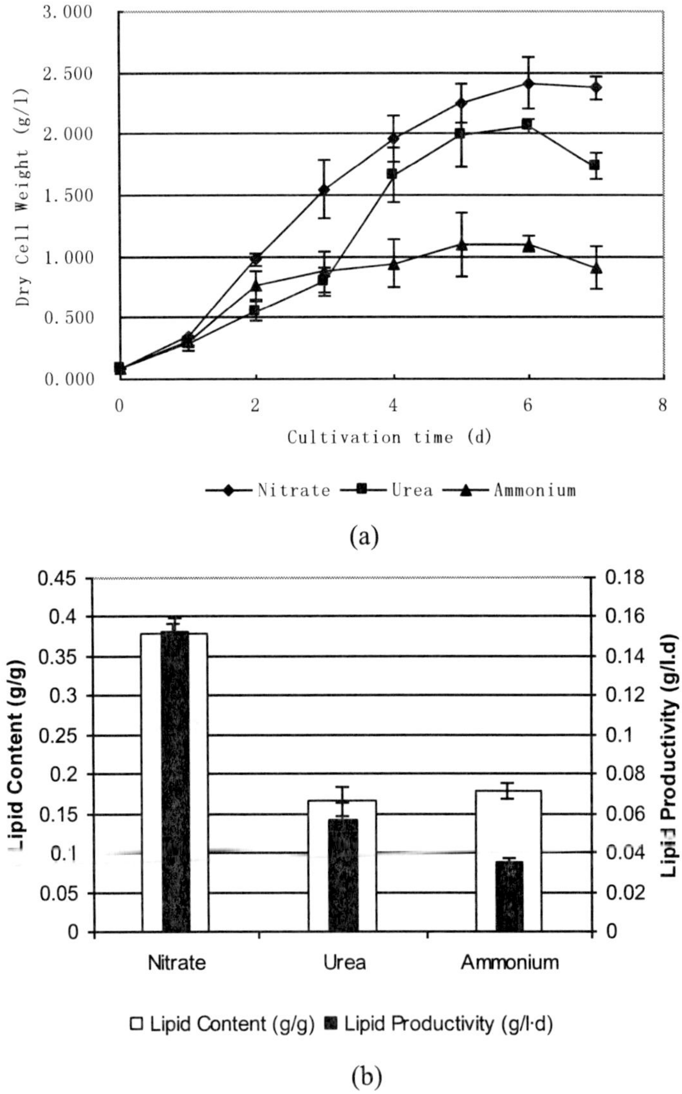

Figure 7.1. (a), (b) Cell growth, lipid contents and lipid productivity of green algae *N. oleoabundans* cultivated in media consisting of 5 mM sodium nitrate, urea or ammonium bicarbonate. Experiments were carried out with triplets.

Table 7.1. Summary of cell growth, lipid production and nitrogen consumption results obtained with media with different sodium nitrate concentrations

Sodium Nitrate Concentration	3 mM	5 mM	10 mM	15 mM	20 mM
Initial $NaNO_3$ (mM)	3.23	4.96	10.29	15.58	20.94
Residual $NaNO_3$ (mM)	0	0	0	0.10	4.97
Maximum Biomass Concentration (g/l)	1.85	2.37	3.15	2.91	2.70
Biomass Productivity (g DCW/l·d)	0.31	0.40	0.63	0.58	0.54
Biomass Yield (g DCW/g N)	40.91	34.13	21.87	13.43	12.08
Lipid Productivity (g/l·d)	0.125	0.133	0.098	0.044	0.038

Figure 7.2(b) shows the lipid cell contents (g/g) in dry cells, total lipid concentration, and lipid productivity obtained at different sodium nitrate concentrations. It is obvious that the lipid cell content decreased when the concentration of sodium nitrate increased in the range of 3 mM to 20 mM. The lipid content analysis also revealed that the algae did not accumulate significant lipids at sodium nitrate concentrations beyond 10 mM and, as aforementioned, the largest lipid productivity occurred when sodium nitrate concentration in the medium was 5 mM. These results suggest that the optimal trade-off between maximizing biomass and lipid content occurs at 5 mM among the tested conditions.

The time course profiles of the chlorophyll contents of cells in cultures with different initial nitrate concentrations are shown in Figure 7.3. Chlorophyll cell content in 3 mM and 5 mM media dropped sharply after the first day of cultivation and a similar decrease occurred in the 10 mM medium at the third day of cultivation. On the other hand, no sharp drop of chlorophyll cell content was observed when the initial sodium nitrate was 15 and 20 mM.

Figure 7.4 shows the time-course profiles of the concentration of sodium nitrate in the cultures with different initial concentrations. As shown in Figure 7.4, the nitrate in media with 3 mM, 5 mM, and 10 mM initial sodium nitrate was depleted after 1, 2, and 3 days, respectively. Not surprisingly, however, substantial residual nitrate was detected at the end of cultivation in media containing 15 and 20 mM nitrate initially.

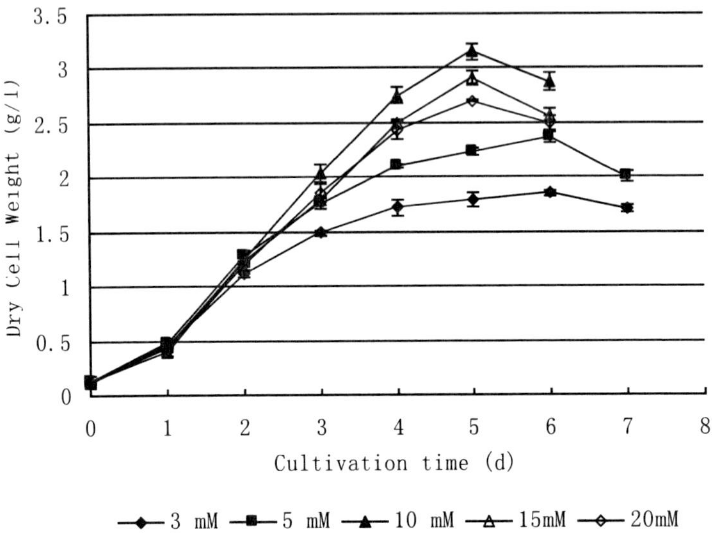

(a)

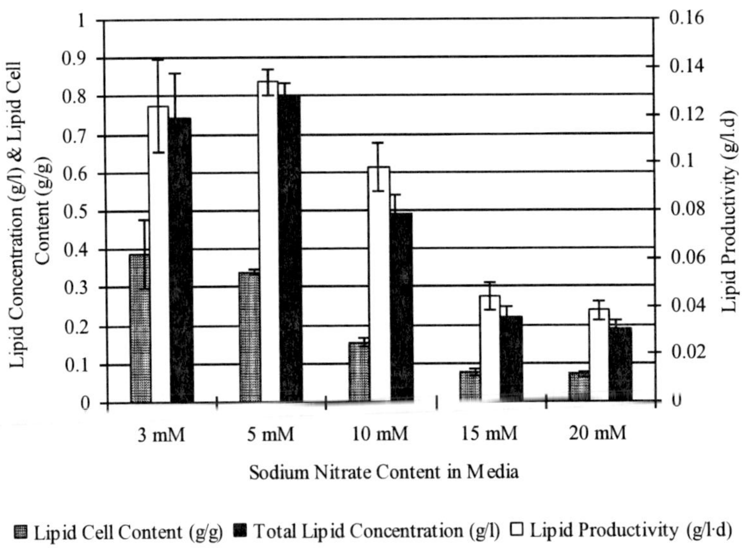

(b)

Figure 7.2. (a), (b) Cell growth, lipid cell contents, total lipid concentration, and lipid productivity of cultivation broth with different sodium nitrate concentrations. Total lipid concentration was estimated by multiplying lipid cell content with corresponding maximum biomass concentration. The lipid productivity was calculated by dividing the total lipid concentration with the number of days of cultivation.

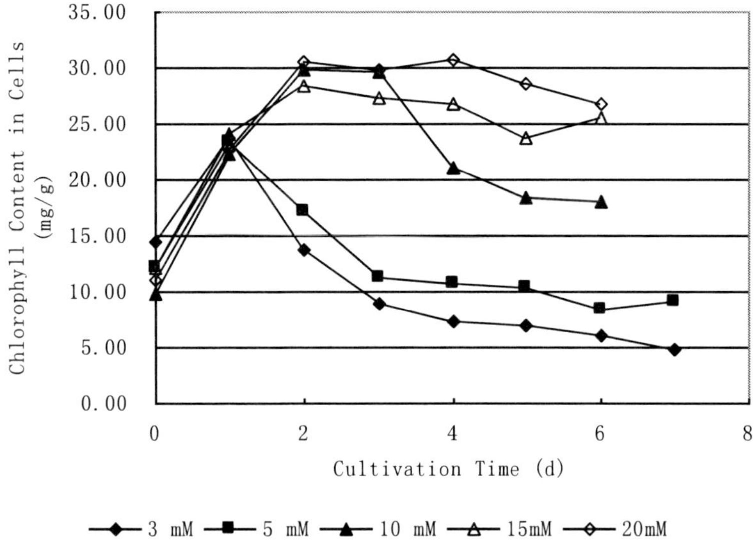

Figure 7.3. Cell growth of *N. oleoabundans* in media of different sodium nitrate.

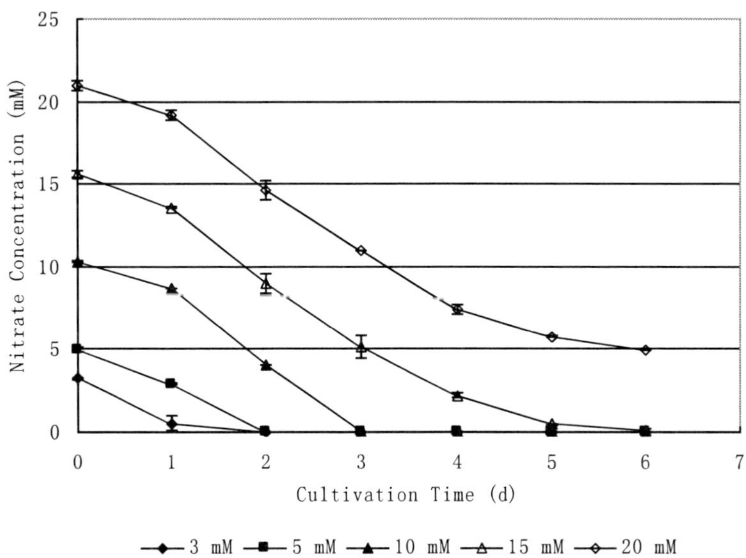

Figure 7.4. Nitrogen consumption of *N. oleoabundans* in the media of different sodium nitrate.

7.4. DISCUSSION

The results show clearly that sodium nitrate was the most favourable nitrogen source for the cell growth and lipid production of N. oleoabundans among the three tested nitrogen-compounds, i.e., sodium nitrate, urea, and ammonium bicarbonate, under the tested conditions. Sodium nitrate was selected to represent nitrate because it is the nitrogen source in the SE medium, which is recommended by the algael collection at the University of Texas, Austin. Besides, it is less costly than potassium nitrate, which would be an important advantage in industrial processes. Ammonium bicarbonate was used to represent ammonia to avoid introducing other anions that could have uncertain effects because the carbonate produced as a result of ammonia consumption would be integrated into the overall inorganic carbon balance in the culture (including dissolved CO2, bicarbonate, and carbonate). However, it would be interesting to further investigate other nitrogen sources such as potassium nitrate, ammonium chloride, and ammonium nitrate on cell growth and lipid production of this algae.

Results also suggest that the optimal sodium nitrate concentration for cell growth was around 10 mM but that supported highest lipid productivity was approximately 5 mM while the lowest sodium nitrate (3 mM) led to the highest lipid cell content (40%). However, sodium nitrate of 15 mM or higher seems to be inhibitive to cell growth. A biomass concentration of 3.15 g/l and a biomass productivity of 0.63 g/l·d were obtained when the nitrate concentration was 10 mM and the highest lipid productivity, which was achieved at 5 mM sodium nitrate, was 0.133 g/l·d. Both these results were remarkable in comparison with previous results, some of which are summarized in Table 7.2. As shown in Table 7.2, the lipid productivity of N. oleoabundans obtained in this study (i.e., 133 mg/l/d) was the highest in comparison to those obtained with phototrophic cultivation of different microalgael species reported in the literature. It was almost two times of the second highest lipid productivity, which was obtained with the phototrophic cultivation of Nannochloris sp. UTEX LB1999 (76.5 mg/l.d) (Yamaberi et al. 1998). It is worth noting that an unusually high lipid productivity (1214 mg/l.d) of the heterotrophic cultivation of Chlorella protothecoides was reported by Xiong et al. (2008). However, the heterotrophic fermentation of microalgae consumes sugars such as glucose and releases CO_2, which does not fulfill the mandate of solar energy capturing and CO_2 mitigation. The high lipid productivity of N. oleoabundans under phototrophic conditions, in combination with the previous results that the lipids of N. oleoabundans were mostly (up to 80%) saturated triglycerides (Tornabene et al. 1983), indicate that this species is a very promising source of triglycerides for biodiesel production.

Table 7.2. Lipid productivity of some microalgael species

Species	P_{DCW} (g/l)/d	Lipid (%)	TAG (%)	T (°C)	P_{lipid} (mg/l·d)	Refs
Parietochloris incise	(7)/14		43-77	25	17.9	(Solovchenko et al. 2008)
Parietochloris incise	(6.3)/38			26	31.5	(Cheng-Wu et al. 2002)
Nannochloris sp. UTEX LB1999	(2.7)/12	34.0	18.8	28	76.5	(Yamaberi et al. 1998)
Nannochloris sp. UTEX LB1999	(2.16)/	50.9	47.6			(Takagi et al. 2000)
Chlorella protothecoides *	(16.8)/8	57.8	NA		1214	(Xiong et al. 2008)
Chlorella emersonii	(1.11)/14	63	NA		50	(Illman et al. 2000)
Chlorella	(0.46)/14	57	NA		18.7	(Illman et al. 2000)
Chlorella vulgaris	(0.52)/14	40	NA		14.9	(Illman et al. 2000)
Chlorella vulgaris	NA	56.6	NA	20	NA	(Liu et al. 2008)
Dunaliella	(0.5)/10	67	NA	28	33.5	(Takagi et al. 2006)
Neochloris oleoabundans	NA	35-54	80	28	NA	(Tornabene et al. 1983)
Neochloris oleoabundans	(2.4)/6	34	NA	34	134	Current Study

* Heterotrophic cultivation

As shown in Figures 2a and 2b and Table 7.1, both the maximum biomass concentration and biomass productivity reached the highest values at 10 mM. These results suggest that, under the investigated conditions, nitrogen was limiting at an initial nitrate concentration lower than a critical value, beyond which the nitrogen limitation was lifted. Considering the fact that, as shown in Figure 7.3, cell chlorophyll content decrease occurred at 10 mM but not at 15 mM nitrate, it seems to be reasonable to hypothesize that this critical value located between 10 mM and 15 mM.

The factor(s) that led to the cessation of cell growth at initial nitrate concentration of 15 mM and 20 mM, while we know for certain was not nitrate depletion, is not clear at this point. Given the high cell density obtained in these experiments, it is quite natural for us to assume that light limitation could be the factor. However, since light was continuously provided at a constant intensity during the cultivation period, light limitation should have led to a linear cell growth rather than cell growth cessation. Cell growth data were collected on a daily basis in these studies, which did not allow accurate analysis of cell growth kinetics. However, it seems to be reasonable to conclude according to the data shown in Figure 7.2a that the algae entered linear growth after the second day of cultivation (biomass concentration at approximately 0.5 OD_{600} for all cultures). The linear growth of algae was particularly obvious for cultivations with high nitrate cultivations, i.e., at 10, 15, and 20 mM nitrate (it was short for low nitrate cultivations such as at 3 and 5 mM). Therefore, light limitation was definitely one of the factors but may not be the sole factor that caused cell growth to enter the stationary phase at high nitrate concentration. It was also observed that cell growth decreased when sodium nitrate further increased from 10 mM to 15 mM and 20 mM. This suggests that sodium nitrate might be inhibitive at these high concentrations.

Figure 7.2(b) shows that the lipid cell content decreased with the concentration of sodium nitrate, the sole nitrogen source in the medium. This observation is compatible with previous research results. For instance, with the same species, Tornabene (1983) achieved a higher lipid content of 54% with a lower sodium nitrate concentration of 0.5 mM sodium nitrate. In the scenario of nutrient-starvation stimulated lipid accumulation, microalgae accumulate lipids as a means of energy storage. The deficiencies of a few nutrients, including phosphate (Khozin-Goldberg and Cohen 2006), sulphur (Sugimoto et al. 2008) and iron (Liu et al. 2008), have been reported as being able to cause the cessation of cell growth and channel metabolic flux generated in photobiosynthesis to lipid/fatty acid biosynthesis and nitrogen is the most commonly reported nutritional limiting factor triggering lipid accumulation in microalgael cells (Khozin-Goldberg et al. 2002; Khozin-Goldberg et al. 2005; Merzlyak et al. 2007; Takagi et al. 2006; Tornabene et al. 1983). The trend of lower sodium nitrate concentration leads to higher lipid cell content could therefore be explained as follows: when the initial nitrogen concentration in medium was low, it would have been exhausted early on at low cell density, when light had good penetration and individual cells would have been exposed to a large quantity of light energy, resulting in more metabolic flux generated from photosynthesis to be channelled to lipid accumulation on an unit biomass basis.

Comparing Figure 7.2(b) and 4, it is apparent that chlorophyll cell content was tightly coupled with the accessibility of cells to external nitrogen sources because the cell content started to drop when the nitrate was exhausted in 3, 5, and 10 mM sodium nitrate media while no significant chlorophyll cell content decrease was observed with 15 and 20 mM sodium nitrate media. Since chlorophyll is a nitrogen-rich compound that contains four nitrogen atoms each molecule, we hypothesize that, when the nitrogen in medium was depleted, cell started to utilize the intracellular nitrogen pool to support the synthesis of cell material for further cell division. It seems that chlorophyll was one of the most easily accessible intracellular nitrogen pools and cells started to utilize them when the external nitrogen pool was depleted. It is worth noticing that, however, chlorophyll is the essential component of the photosynthetic mechanism of green algae, which is responsible for capturing solar energy and CO_2 to generate the metabolic flux for both cell growth and lipid accumulation. Cell growth and lipid accumulation would therefore be impeded if the chlorophyll cell content drops to below a critical level.

Comparing Figures 2(a) and 4, it is obvious that cell growth continued after nitrogen source was exhausted from the culture when sodium nitrate concentration was 3, 5 or 10 mM. As discussed previously, intracellular nitrogen pools such as chlorophyll molecules were likely to be consumed to support the cell growth after nitrogen exhaustion. This is supported by the observation on the sharp drop of chlorophyll cell content of algael cells (Figure 7.3), which coincided with the nitrate depletion from the corresponding cultures (Figure 7.4). This observation is compatible with the results obtained with the media containing 15 mM and 20 mM media, which showed neither nitrate depletion nor sharp drop of chlorophyll content in the whole course of cultivation.

The above observation can be illustrated more clearly by Figure 7.5, which shows the profiles of biomass concentration, chlorophyll cell content and nitrate of a typical culture with 5 mM sodium nitrate. As shown in Figure 7.5, sodium nitrate, the sole nitrogen source in the medium, was depleted at the second day of cultivation. However, the biomass concentration continued to increase until the sixth day of cultivation with biomass concentration increased from approximately 1.2 g/l to 2.4 g/l. In other words, biomass concentration doubled after the external nitrogen source was exhausted. Apparently, intracellular nitrogen-rich compounds must have been converted to essential nitrogen-containing cell materials such as proteins, nucleic acid, and cell wall materials to support the cell growth.

As also shown in Figure 7.5, the chlorophyll cell content increased from around 3 mg/g DCW at time 0 (the initial biomass from inoculum) to approximately 5.6 mg/g DCW at day one, and then decreased drastically to approximately 2.7 mg/g DCW at day three. This observation suggests that cells accumulated large quantities of chlorophyll molecules when nitrogen source was abundantly available. Then, when the external nitrogen sources are exhausted, cells started to utilize chlorophylls as an intracellular nitrogen source.

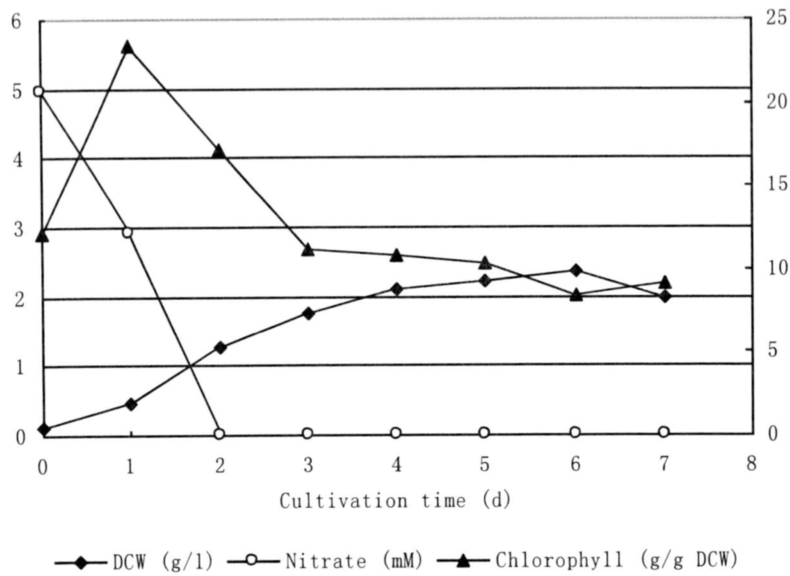

Figure 7.5. Typical time-course profiles of N. oleoabundans cultivated in medium containing 5 mM sodium nitrate.

ACKNOWLEDGMENT

Financial supports from Natural Science and Engineering Research Council of Canada (NSERC) and Canadian Foundation of Innovation (CFI) are gratefully acknowledged.

REFERENCES

Barnhart MC. 2006. Buckets of muckets: A compact system for rearing juvenile freshwater mussels. *Aquaculture* 254(1-4):227-233.

Becker EW. 1994. *Microalgae: biotechnology and microbiology*. NY: Cambridge University Press.

Bigogno C, Khozin-Goldberg I, Boussiba S, Vonshak A, Cohen Z. 2002. Lipid and fatty acid composition of the green oleaginous algae Parietochloris incisa, the richest plant source of arachidonic acid. *Phytochemistry* 60(5):497-503.

Cheng-Wu Z, Cohen Z, Khozin-Goldberg I, Richmond A. 2002. Characterization of growth and arachidonic acid production of Parietochloris incisa comb. nov

(Trebouxiophyceae, Chlorophyta). *Journal of Applied Phycology* 14(6):453-460.

Chisti Y. 2007. Biodiesel from microalgae. *Biotechnology Advances* 25(3):294-306.

Gatenby CM, Orcutt DM, Kreeger DA, Parker BC, Jones VA, Neves RJ. 2003. Biochemical composition of three algael species proposed as food for captive freshwater mussels. *Journal of Applied Phycology* 15(1):1-11.

Gatenby CM, Parker BC, Neves RJ. 1997. Growth and survival of juvenile rainbow mussels, Villosa iris (Lea, 1829) (Bivalvia: Unionidae), reared on algael diets and sediment. *American Malacological Bulletin* 14(1):57-66.

Gordillo FJL, Goutx M, Figueroa FL, Niell FX. 1998. Effects of light intensity, CO2 and nitrogen supply on lipid class composition of Dunaliella viridis. *Journal of Applied Phycology* 10(2):135-144.

Illman AM, Scragg AH, Shales SW. 2000. Increase in Chlorella strains calorific values when grown in low nitrogen medium. *Enzyme and Microbial Technology* 27(8):631-635.

Jones JW, Mair RA, Neves RJ. 2005. Factors affecting survival and growth of juvenile freshwater mussels cultured in recirculating aquaculture systems. *North American Journal of Aquaculture* 67(3):210-220.

Jones JW, Neves RJ. 2002. Life history and propagation of the endangered fanshell pearlymussel, Cyprogenia stegaria Rafinesque (Bivalvia: Unionidae). *Journal of the North American Benthological Society* 21(1):76-88.

Jovanovic M, Djukic M, Vasiljevic I, Ninkovic M, Jovanovic M. 2007. Determination of nitrate by the IE-HPLC-UV method in the brain tissues ofWistar rats poisoned with paraquat. *J. Serb. Chem. Soc.* 72(4):347–356.

Khozin-Goldberg I, Bigogno C, Shrestha P, Cohen Z. 2002. Nitrogen starvation induces the accumulation of arachidonic acid in the freshwater green algae Parietochloris incisa (Trebuxiophyceae). *Journal of Phycology* 38(5):991-994.

Khozin-Goldberg I, Cohen Z. 2006. The effect of phosphate starvation on the lipid and fatty acid composition of the fresh water eustigmatophyte Monodus subterraneus. *Phytochemistry* 67(7):696-701.

Khozin-Goldberg I, Shrestha P, Cohen Z. 2005. Mobilization of arachidonyl moieties from triacylglycerols into chloroplastic lipids following recovery from nitrogen starvation of the microalgae Parietochloris incisa. *Biochimica et Biophysica Acta - Molecular and Cell Biology of Lipids* 1738(1-3):63-71.

Li Y, Qin JG. 2005. Comparison of growth and lipid content in three Botryococcus braunii strains. Journal of Applied Phycology 17(6):551-556.

Li Y. HM, Wu N, Lan CQ, Dubois-Calero N. 2008. *Biofuels from Microalgae. Biotech Prog*. in press.

Liu ZY, Wang GC, Zhou BC. 2008. Effect of iron on growth and lipid accumulation in Chlorella vulgaris. *Bioresource Technology* 99(11):4717-4722.

Merzlyak MN, Chivkunova OB, Gorelova OA, Reshetnikova IV, Solovchenko AE, Khozin-Goldberg I, Cohen Z. 2007. Effect of nitrogen starvation on optical properties, pigments, and arachidonic acid content of the unicellular green algae parietochloris incisa (Trebouxiophyceae, Chlorophyta). *Journal of Phycology* 43(4):833-843.

Metzger P, Largeau C. 2005. Botryococcus braunii: A rich source for hydrocarbons and related ether lipids. *Applied Microbiology and Biotechnology* 66(5):486-496.

Solovchenko AE, Khozin-Goldberg I, Didi-Cohen S, Cohen Z, Merzlyak MN. 2008. Effects of light intensity and nitrogen starvation on growth, total fatty acids and arachidonic acid in the green microalgae Parietochloris incisa. *Journal of Applied Phycology* 20(3):245-251.

Sugimoto K, Midorikawa T, Tsuzuki M, Sato N. 2008. Upregulation of PG synthesis on sulfur-starvation for PS I in Chlamydomonas. *Biochemical and Biophysical Research Communications* 369(2):660-665.

Takagi M, Karseno, Yoshida T. 2006. Effect of salt concentration on intracellular accumulation of lipids and triacylglyceride in marine microalgae Dunaliella cells. *Journal of Bioscience and Bioengineering* 101(3):223-226.

Takagi M, Watanabe K, Yamaberi K, Yoshida T. 2000. Limited feeding of potassium nitrate for intracellular lipid and triglyceride accumulation of Nannochloris sp. UTEX LB1999. *Applied Microbiology and Biotechnology* 54(1):112-117.

Tornabene TG, Holzer G, Lien S, Burris N. 1983. Lipid composition of the nitrogen starved green algae Neochloris oleoabundans. *Enzyme and Microbial Technology* 5(6):435-440.

Usui N, Ikenouchi M. 1997. The biological CO2 fixation and utilization project by RITE(1): Highly-effective photobioreactor system. *Energy Conversion and Management* 38(SUPPL. 1):S487-S492.

Xiong W, Li X, Xiang J, Wu Q. 2008. High-density fermentation of microalgae Chlorella protothecoides in bioreactor for microbio-diesel production. *Applied Microbiology and Biotechnology* 78(1):29-36.

Xu H, Miao X, Wu Q. 2006. High quality biodiesel production from a microalgae Chlorella protothecoides by heterotrophic growth in fermenters. *Journal of Biotechnology* 126(4):499-507.

Yamaberi K, Takagi M, Yoshida T. 1998. Nitrogen depletion for intracellular triglyceride accumulation to enhance liquefaction yield of marine microalgael cells into a fuel oil. *Journal of Marine Biotechnology* 6(1):44-48.

INDEX

B

C

D

E

M

P

Q

R

S

V

W

X

Y

Z